中 外 物 理 学 精 品 书 系
本书出版得到"国家出版基金"资助

中外物理学精品书系

引进系列·4

Dust in the Universe:
Similarities and Differences

宇宙尘埃
——相似性和差异性

(影印版)

〔印〕斯瓦米(K. S. K. Swamy) 著

著作权合同登记号　图字：01-2012-2819

图书在版编目(CIP)数据

Dust in the Universe：Similarities and Differences ＝ 宇宙尘埃：相似性和差异性：英文/(印)斯瓦米(Swamy，K. S. K.)著. —影印本. —北京：北京大学出版社，2012.12

(中外物理学精品书系·引进系列)

ISBN 978-7-301-21660-6

Ⅰ. ①宇… Ⅱ. ①斯… Ⅲ. ①星际尘埃-研究-英文 Ⅳ. ①P155.2

中国版本图书馆 CIP 数据核字(2012)第 282013 号

Copyright © 2005 by World Scientific Co. Pte. Ltd. All rights reserved. This book, or parts thereof, may not be reproduced in any form or by any means, electronic or mechanical, including photocopying, recording or any information storage and retrieval system now known or to be invented, without written permission from the Publisher.

Reprint arranged with World Scientific Co. Pte. Ltd., Singapore.

书　　名：Dust in the Universe: Similarities and Differences(宇宙尘埃——相似性和差异性)(影印版)

著作责任者：〔印〕斯瓦米(K. S. K. Swamy) 著

责 任 编 辑：刘　啸

标 准 书 号：ISBN 978-7-301-21660-6/O·0903

出 版 发 行：北京大学出版社

地　　　址：北京市海淀区成府路 205 号　100871

网　　　址：http://www.pup.cn

新 浪 微 博：@北京大学出版社

电 子 信 箱：zpup@pup.cn

电　　　话：邮购部 62752015　发行部 62750672　编辑部 62752038
　　　　　　出版部 62754962

印 刷 者：北京中科印刷有限公司

经 销 者：新华书店
　　　　　　730 毫米×980 毫米　16 开本　16.75 印张　310 千字
　　　　　　2012 年 12 月第 1 版　2012 年 12 月第 1 次印刷

定　　　价：45.00 元

未经许可，不得以任何方式复制或抄袭本书之部分或全部内容。

版权所有，侵权必究

举报电话：010-62752024　电子信箱：fd@pup.pku.edu.cn

《中外物理学精品书系》
编委会

主　任：王恩哥
副主任：夏建白
编　委：(按姓氏笔画排序，标 * 号者为执行编委)

王力军	王孝群	王　牧	王鼎盛	石　兢
田光善	冯世平	邢定钰	朱邦芬	朱　星
向　涛	刘　川*	许宁生	许京军	张　酣*
张富春	陈志坚*	林海青	欧阳钟灿	周月梅*
郑春开*	赵光达	聂玉昕	徐仁新*	郭　卫*
资　剑	龚旗煌	崔　田	阎守胜	谢心澄
解士杰	解思深	潘建伟		

秘　书：陈小红

序　言

物理学是研究物质、能量以及它们之间相互作用的科学。她不仅是化学、生命、材料、信息、能源和环境等相关学科的基础，同时还是许多新兴学科和交叉学科的前沿。在科技发展日新月异和国际竞争日趋激烈的今天，物理学不仅囿于基础科学和技术应用研究的范畴，而且在社会发展与人类进步的历史进程中发挥着越来越关键的作用。

我们欣喜地看到，改革开放三十多年来，随着中国政治、经济、教育、文化等领域各项事业的持续稳定发展，我国物理学取得了跨越式的进步，做出了很多为世界瞩目的研究成果。今日的中国物理正在经历一个历史上少有的黄金时代。

在我国物理学科快速发展的背景下，近年来物理学相关书籍也呈现百花齐放的良好态势，在知识传承、学术交流、人才培养等方面发挥着无可替代的作用。从另一方面看，尽管国内各出版社相继推出了一些质量很高的物理教材和图书，但系统总结物理学各门类知识和发展，深入浅出地介绍其与现代科学技术之间的渊源，并针对不同层次的读者提供有价值的教材和研究参考，仍是我国科学传播与出版界面临的一个极富挑战性的课题。

为有力推动我国物理学研究、加快相关学科的建设与发展，特别是展现近年来中国物理学者的研究水平和成果，北京大学出版社在国家出版基金的支持下推出了《中外物理学精品书系》，试图对以上难题进行大胆的尝试和探索。该书系编委会集结了数十位来自内地和香港顶尖高校及科研院所的知名专家学者。他们都是目前该领域十分活跃的专家，确保了整套丛书的权威性和前瞻性。

这套书系内容丰富，涵盖面广，可读性强，其中既有对我国传统物理学发展的梳理和总结，也有对正在蓬勃发展的物理学前沿的全面展示；既引进和介绍了世界物理学研究的发展动态，也面向国际主流领域传播中国物理的优秀专著。可以说，《中外物理学精品书系》力图完整呈现近现代世界和中国物理科学发展的全貌,是一部目前国内为数不多的兼具学术价值和阅读乐趣的经典物理丛书。

《中外物理学精品书系》另一个突出特点是,在把西方物理的精华要义"请进来"的同时,也将我国近现代物理的优秀成果"送出去"。物理学科在世界范围内的重要性不言而喻,引进和翻译世界物理的经典著作和前沿动态,可以满足当前国内物理教学和科研工作的迫切需求。另一方面,改革开放几十年来,我国的物理学研究取得了长足发展,一大批具有较高学术价值的著作相继问世。这套丛书首次将一些中国物理学者的优秀论著以英文版的形式直接推向国际相关研究的主流领域,使世界对中国物理学的过去和现状有更多的深入了解,不仅充分展示出中国物理学研究和积累的"硬实力",也向世界主动传播我国科技文化领域不断创新的"软实力",对全面提升中国科学、教育和文化领域的国际形象起到重要的促进作用。

值得一提的是,《中外物理学精品书系》还对中国近现代物理学科的经典著作进行了全面收录。20世纪以来,中国物理界诞生了很多经典作品,但当时大都分散出版,如今很多代表性的作品已经淹没在浩瀚的图书海洋中,读者们对这些论著也都是"只闻其声,未见其真"。该书系的编者们在这方面下了很大工夫,对中国物理学科不同时期、不同分支的经典著作进行了系统的整理和收录。这项工作具有非常重要的学术意义和社会价值,不仅可以很好地保护和传承我国物理学的经典文献,充分发挥其应有的传世育人的作用,更能使广大物理学人和青年学子切身体会我国物理学研究的发展脉络和优良传统,真正领悟到老一辈科学家严谨求实、追求卓越、博大精深的治学之美。

温家宝总理在2006年中国科学技术大会上指出,"加强基础研究是提升国家创新能力、积累智力资本的重要途径,是我国跻身世界科技强国的必要条件"。中国的发展在于创新,而基础研究正是一切创新的根本和源泉。我相信,这套《中外物理学精品书系》的出版,不仅可以使所有热爱和研究物理学的人们从中获取思维的启迪、智力的挑战和阅读的乐趣,也将进一步推动其他相关基础科学更好更快地发展,为我国今后的科技创新和社会进步做出应有的贡献。

《中外物理学精品书系》编委会　主任
中国科学院院士,北京大学教授
王恩哥
2010年5月于燕园

World Scientific Series in Astronomy and Astrophysics – Vol. 7

Dust in the Universe

Similarities and Differences

K S Krishna Swamy

National Astronomical Observatory, Japan

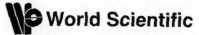

NEW JERSEY • LONDON • SINGAPORE • BEIJING • SHANGHAI • HONG KONG • TAIPEI • CHENNAI

Preface

We live in a dusty Universe. Dust plays an important role in the overall scenario of structure and evolution of the universe. Observations carried out with the Infrared Space Observatory showed for the first time highly complex mineralogy of the dust material and its variation among different astronomical objects in the universe, leading perhaps to a new area of astromineralogy. Extensive and high quality observations carried out using ground-based telescopes and satellites have led to phenomenal progress in our understanding of the nature and composition of dust in the universe upto redshifts $z \sim 6.4$. This shows that dust is present from almost the beginning of the Universe to the present time. The method of studying the evolution of dust is in the reverse order; starting with local universe and then moving farther in space which means going back to earlier times. Thus it seems that this is the appropriate time to summarize the present status of cosmic dust, essentially evolution of astromineralogy.

I would like to thank G. Wallerstein for impressing upon me the importance of such a book and encouraging me to take up the venture. It is a pleasure to thank Juni-ichi Watanabe and the National Astronomical Observatory of Japan for providing excellent research enviornment leading to this book. I am grateful to J.V. Narlikar for encouragement in the publication of the book. I am also thankful to Tata Institute of Fundamental Research for the use of their excellent facilities.

I am especially grateful to my colleagues H.M. Antia who helped me in various technical ways, from literature survey to digitization and S. Ramadurai for a thorough reading of the manuscript. I would also like to express my gratitude to my parents and my family for their support and encouragements. I would like to thank all the publishers and authors, who have given permission for reproduction of figures and tables. The help of

M. Shinde in the preparation of the manuscript is greatly appreciated.

204 Sigma Tower   K.S.Krishna Swamy
Plot 32, Gorai Road
Borivli (W), Mumbai 400 091
India
January 2005

Contents

Preface   v

1. General Introduction   1
       References 4

2. Laboratory Studies   5
      2.1 Introduction 5
      2.2 Infrared Spectroscopy 6
      2.3 Mixed Ices 10
      2.4 Silicate 17
          2.4.1 Structure of Silicate 17
          2.4.2 Silicate Studies 17
              2.4.2.1 Condensation 19
              2.4.2.2 Effect of Environment 20
              2.4.2.3 Silicon Nano-crystals 22
              2.4.2.4 Optical Constants 23
      2.5 Carbon 23
          2.5.1 Forms of Carbon 23
          2.5.2 Carbon Studies 25
      2.6 Polycyclic Aromatic Hydrocarbons (PAHs) 27
          2.6.1 Spectra 27
          2.6.2 Mass Spectra 33
          2.6.3 Steller Environment 33
          2.6.4 Diffuse Interstellar Bands 37
      2.7 Optical Properties of Materials 37
          2.7.1 Theoretical Considerations 37

2.7.2 Laboratory Measurements	38
2.7.3 Mass Absorption Coefficient	41
2.7.4 Microwave Scattering	44
2.8 Microgravity Studies	45
2.9 Nucleation	48
2.10 Coagulation and Aggregation	49
2.11 Other Dust Studies	52
References	52

3. **Interstellar Dust** — 57

3.1 Introduction	57
3.2 Estimate of Amount of Extinction	59
3.3 Effect on Derived Distances	62
3.4 Amount of Absorbing Material	63
3.5 Nature of Dust	64
3.5.1 Mean Interstellar Reddening Curve	64
3.5.2 Theoretical Extinction Curve	68
3.5.3 Variations in the Interstellar Extinction Curve	71
3.6 Interstellar Polarization	72
3.7 Scattered Light	73
3.7.1 Diffuse Galactic Light	73
3.7.2 Reflection Nebulae	75
3.7.3 The Extended Red Emission	77
3.8 Elemental Depletion	79
3.9 Diffuse Interstellar Bands	81
3.10 Infrared Spectral Features	82
3.10.1 Diffuse Interstellar Medium	82
3.10.2 HII Region	84
3.10.3 Reflection Nebulae	85
3.10.4 Molecular Clouds	86
3.10.5 Processes in Molecular Clouds	92
3.10.6 Young Massive Protostar	93
3.10.7 Star-Forming Regions	94
3.11 Galactic Centre	96
3.11.1 Extinction Law	99
3.12 Sources of Dust	101
3.13 Detection of Interstellar Dust: *in-situ*	103
3.13.1 Spacecraft Studies	103
3.13.2 Presolar Grains	105

References 107

4. Cometary Dust 111

    4.1 Introduction 111
    4.2 Dust Tails 114
        4.2.1 Dynamics of Dust Tails 114
        4.2.2 Dust Trail 116
    4.3 Radiation Pressure Effects 117
    4.4 Visible Continuum 118
    4.5 Phase Function 119
    4.6 Polarization 120
    4.7 Infrared Observations 122
    4.8 Albedo 124
    4.9 Spectral Features in the Infrared 126
        4.9.1 Spectral Features 126
        4.9.2 Sizes of Grains 130
    4.10 Organics 133
    4.11 Water-Ice 137
    4.12 Mineralogical Composition 138
    4.13 Isotopic Studies 141
    References 143

5. Interplanetary Dust 147

    5.1 Introduction 147
    5.2 Interplanetary Dust Particles 147
        5.2.1 Morphology, Structure and Chemical Composition . 147
        5.2.2 Origin 153
    5.3 Meteorites 156
        5.3.1 Morphology, Structure and Chemical Composition . 157
        5.3.2 Presolar Grains 159
        5.3.3 Time of Formation 163
    5.4 Extraterrestrial Origin 164
        5.4.1 Presolar Grains 165
    References 167

6. Circumstellar Dust 169

    6.1 Introduction 169
    6.2 AGB Stars 169

    6.3 Mass Loss from Stars 170
    6.4 Theoretical Considerations 172
          6.4.1 Dust Formation 172
          6.4.2 Condensation of Dust 173
                6.4.2.1 C/O < 1 (Oxygen-rich stars) 173
                6.4.2.2 C/O >1 (Carbon-rich stars) 176
                6.4.2.3 C/O = 1 (S-Stars) 177
          6.4.3 Circumstellar Chemistry 178
    6.5 Observational Results 180
          6.5.1 Carbon-rich Stars 180
          6.5.2 Oxygen-rich Stars 182
          6.5.3 S Stars 186
          6.5.4 UIR Bands in Oxygen-rich Supergiants 186
          6.5.5 Stars of Other Type 187
                6.5.5.1 Herbig Ae/Be Stars 188
                6.5.5.2 Vega-Type Stars 191
          6.5.6 Planetary Nebulae 192
          6.5.7 Novae 195
          6.5.8 Supernovae 197
          References 199

7. Extragalactic Dust   203

    7.1 Introduction 203
    7.2 Magellanic Cloud 205
          7.2.1 Extinction Curve 206
          7.2.2 Spectral Features 208
    7.3 Normal Galaxies 209
    7.4 Seyfert Galaxies 213
    7.5 Starburst Galaxies 215
    7.6 Ultraluminous Infrared Galaxies 216
    7.7 Merging Galaxies 217
    7.8 Virgo and Coma Clusters 218
    7.9 Quasars 219
          7.9.1 Far-Infrared Radiation 219
          7.9.2 Spectral Studies 221
                7.9.2.1 Quasar Environment 222
                7.9.2.2 Lyman α Forest 223
                7.9.2.3 Damped Lyman α Systems 224
    7.10 Intergalactic Dust 225

    7.11 Intracluster Medium 226
    7.12 Cosmic Background Radiation 228
        7.12.1 Cosmic Infrared Background Radiation 228
        7.12.2 Cosmic Microwave Background Radiation 230
        References 231

8. Epilogue   235

Appendix A   H-R Diagram   239

Appendix B   Stellar Evolution   241

Appendix C   Nucleosynthesis   243

Appendix D   Acronyms   245

Index   249

Chapter 1

General Introduction

Study of the dust particles in the universe is an exciting and challenging area of current astronomy. This comes from various considerations. Dust is at the centre of all important events in the universe, like star formation, distance determination etc. Dust particles influence the thermal, dynamical, ionization and chemical states of matter of the interestellar and intergalactic medium. It may be noted, about half of the elements heavier than helium is in the form of dust in the interstellar medium of our Galaxy. Dust grains are effective absorbers of radiation energy and momentum. The collisional coupling between grains and gas in dense regions around evolved stars, in molecular cloud cores and protoplanetary disks have consequences for the temperature and the dynamical evolution of the gas component. The mass loss from evolved stars, stability of molecular clouds and their fragmentation into smaller clumps etc. are directly linked to the presence of dust. Infrared observations arising due to thermal emission from dust particles probe regions of star formation. The low temperature inside clouds arising from shielding of radiation by the dust particles leads to complex chemistry between dust and atoms and molecules. The formation of molecular hydrogen takes place on the surface of grains. The dust and complex chemistry of the material present in comets appear to be the same as in these clouds. May be these are the precursors to life on earth! However due to historical reasons and also due to lack of facilities both with regard to bigger telescopes and technology, most of the studies had to be concentrated on the local universe, particularly our Galaxy and nearby ones. Even here, until recently, studies were concentrated on brighter objects. Things have improved enormously with the advent of space and new technology telescopes. It is now possible to push the limit of observations to fainter and fainter sources. Also it has been possible to study objects farther in space and hence well back in time.

Nature and composition of dust in the Galaxy have come from many kinds of observations such as extinction, scattering and polarization of starlight, infrared continuum emission, spectral features in the infrared region, interstellar depletion of elements, presence of isotopic anomalies in meteorites etc. These studies have been carried out on different kinds of objects such as diffuse and dense clouds, Reflection nebulae, star forming regions, HII regions, circumstellar shells of stars, meteorites, interplanetary dust particles etc. High resolution spectral observations in the infrared region have given enormous information about the composition of dust particles. All these studies have shown that dust particles are mainly made up of crystalline and amorphous silicates and of carbonaceous material in some form. The near and mid-infrared spectra of many sources are dominated by strong emission features at 3.3, 6.2, 7.7, 8.6 and 11.3μm. These are collectively known as unidentified infrared (UIR) bands. However it is suggestive that they could arise from the large sized molecules containing about 50 to 100 atoms, such as Polycyclic Aromatic Hydrocarbon (PAHs). However no specific molecule has been identified and it is further complicated by the fact that emission arises not from a single molecule but from a host of molecules. However the carrier appears to be aromatic in character. In view of this, We prefer to use the term "Aromatic Unidentified Infrared Bands" (AUIBs) to describe this generic spectrum. The grains inside molecular clouds are heavily processed with a mantle of various ices. Nature of the dust is also dependent on the radiation field present in the environment.

The major supply of dust comes from circumstellar shell of evolved stars such as oxygen-rich and carbon-rich stars producing silicate type and carbonaceous type of materials respectively. Stellar winds eject these dust particles into interstellar medium. These particles cycle several times through dense and diffuse clouds which could modify their original composition. The chemical and physical structure of dust grains are also modified by a host of processes including UV photon irradiation, gas-phase chemistry, accretion and grain surface reactions, cosmic ray bombardment and destruction by shock waves generated by supernova explosion etc. Therefore the nature and chemical composition of grains in the Galaxy is a function of the environment and quite complicated.

An understanding of the chemical complexity of the grain material in Galaxy is crucial for an understanding of the chemical state of dust in extragalactic systems and in the universe. It may be noted that the conditions existing in astrophysical objects are diverse and extreme. The temperature, particle density and radiation field can vary over a wide range of values. Therefore the success of the study of dust grains in the Galaxy and in extragalactic systems is critically dependent upon laboratory studies of dust carried out under astrophysical conditions. Extensive laboratory studies carried out with dust analogues have helped enormously our understanding of the chemical composition of dust in various locations. Since the laws of physics and chemistry are the same throughout we can extend the studies of local environment to the universe.

Other galaxies in the universe can have various shapes defined as spirals (our Galaxy), ellipticals etc. which themselves form clusters and superclusters. Dust is present, as we have seen above, not only within the galaxies but in the intergalactic and intracluster media as well. One major difference of course is that chemical composition of the gaseous material in the past is not the same as that of the Galaxy. The earliest heavy elements could not have been created until after the first stars and the first supernova, roughly at redshift $z \sim 6$. This could in principle have an effect on the nature of dust present in these objects. Dust condenses out of the gaseous material. This is the processed material inside stars due to nucleosynthetic processes. The processed gaseous material inside the star is returned to the interstellar medium which enriches the original material. The stars formed out of the new material will be richer in heavier elements compared to the previous generation of stars. Once the dust is formed within the galaxies it can be expelled to extragalactic space by several processes like galactic wind, explosions etc. The modification of the material has been going on through such a cycle over the age of the universe.

As mentioned earlier, dust manifests in different ways. However detailed nature of the dust can be determined from objects which are close by. As one goes out to larger and larger distances objects become fainter and fainter. Therefore one can estimate only some gross property of the grains. This situation should improve in future with the availability of bigger telescopes and sensitive detectors. However at the present time, Quasars, the highly redshifted objects are well suited for this study at large distances as they are very bright. At the present time they have traced the state of the universe upto redshift of $z \sim 6$.

In the succeeding chapters we would like to elaborate on some of these aspects. We will first present some of the Laboratory studies relating to the nature of dust, like silicates, carbonaceous materials etc. This will then be followed by the available information on the nature and composition of dust grains in the universe from a representative sample of various kinds of objects present in the universe from the present epoch to earlier epochs. i.e. first galactic dust and then extragalactic dust. We follow this approach since we have a reasonably good idea on the nature, composition and formation of dust in the Galaxy. In the same spirit we will discuss the galactic dust in a little more detail. In this manner we hope to arrive at the present status of the study of dust in the universe.

References

Following are some of the publications covering the subject of dust.

d'Hendecourt, L., Joblin, C. and Jones, A. (Eds.) 1999, *Solid Interstellar Matter: ISO Revolution*, Springer-Verlag.

Drain, B.T. 2003, Interstellar Grains, Ann. Rev. Astron. Astrophys., **41**, 241.

Ehrenfreund, P., Kraft, C., Kochan, H and Pirronello, V. (Eds.) 1999, *Laboratory Astrophysics and Space Research*, Kluwer Academic Publishers, Dodrecht.

Femiano, R.F. and Matteucci, F. (Eds.) 2002, *Chemical Enrichment of Intracluster and Intergalactic Medium*, ASP Conference Series, Vol. 253, Astronomical Society of the Pacific Publishers.

Greenberg, J.M. (Ed.) 1996, *The Cosmic Dust Connection*, Kluwer, Dordrecht.

Grun, E., Gustafson, Bo. A.S., Dermott, S. and Fechtig, H. (Eds.) 2001, *Interplanetary Dust*, Springer-Verlag, Berlin.

Hoyle, F and Wickramasinghe, N.C. 1991, *The Theory of Cosmic Grains*, Kluwer Academic Publishers, Dordrecht.

Hulst, van der, J.M. (Ed.) 1997, *The Interstellar Matter in Galaxies*, Kluwer Academic Publishers, Dordrecht.

Pendleton, Y.J.and Tielens, A.G.G.M. (Eds.) 1997, *From Stardust to Planetesimals*, ASP Conference Series, Vol. 122, Astronomical Society of the Pacific Publishers.

Winckel, H.V. 2003, Post-AGB Stars, Ann. Rev. Astron. Astrophys., **41**, 391.

Witt, A.N., Clayton, G.C. and Drain, B.T. (Eds.) 2004, *Astrophysics of Dust*, ASP Conference Series, Vol. 309, Astronomical Society of The Pacific Publishers.

Chapter 2

Laboratory Studies

2.1 Introduction

Studies relating to atoms and molecules are very extensive. From astrophysical point of view, some of the most important studies are spectroscopic constants, wavelength of lines, oscillator strengths, collision crosssections, chemical reaction rates etc. These are done under laboratory conditions. But studies pertaining to physical conditions existing in space are very few in number. The situation is still worse with regard to studies pertaining to dust grains. In recent years considerable effort is being excercised to improve the situation. Several groups in the world are actively involved in the various studies relating to dust grains in space.

As pointed out earlier, there is a constant interplay between gas and dust in astrophysical objects. These involve a knowledge of grain properties, gas-dust interaction and grain surface chemistry. There are also grains of various sizes, starting from very small ones, about 50 to 100 atoms or molecules, to large sizes upto hundreds of molecules. The gas and dust are also subjected simultaneously to a host of physical and chemical processes. These include UV photon irradiation, gas-phase chemistry, accretion and grain surface reactions, cosmic ray bombardment, disruption by shock waves generated by supernova explosion and so on. Therefore, the whole process is highly complex. It is rather difficult to take into account all these effects simultaneously in the laboratory studies. So experiments are generally carried out with one or more effects considered. Here we would like to outline some of these studies.

2.2 Infrared Spectroscopy

The most commonly used method of studying the composition of the material is infrared spectroscopy. This has the great advantage that laboratory studies can directly be compared with the astronomical spectra to identify and also determine the abundance of the species.

If the two atoms in a diatomic molecule are of the same type, like H_2, N_2, O_2 etc., it is called Homonuclear molecule. If the two atoms are of different type, like CH, CN, OH etc., they are called Hetronuclear molecule. The spectrum of even the simplest diatomic molecule shows complicated behaviour comprising different bands and each band itself is made up of many lines. This is due to the fact that the two atoms in the molecule can vibrate individually along the main axis as well as rotate along the axis perpendicular to the common axis. The total energy of the molecule is the sum total of kinetic and potential energies of electrons and nucleus. This is generally represented in terms of the potential energy curve, which give the variation of the energy as a function of the internuclear separation. The energy required to separate a stable molecule into its components is called the Dissociation energy of the molecule. The various vibrational energies of the molecule are denoted by the vibrational quantum number v and can take values 0,1,2.... Each vibrational level is further split into various rotational levels denoted by the rotational quantum number J. Therefore, each electronic state has many vibrational levels, each of which in turn has several rotational levels. A transition can take place between vibrational and rotational levels of the two electronic states. The electronic transitions give lines which fall in the visible and ultraviolet regions, vibrational transitions lie in the infrared regions and those of rotational transitions lie in the radio regions. Pure vibrational transitions in a given electronic state are allowed for hetronuclear molecules, but not for homonuclear molecules. The spectra of polyatomic molecules are far more complicated than the spectra of diatomic molecules due to the presence of more atoms and complex interactions among these atoms. In general, large number of lines are present in the spectra. Therefore, at low spectral resolution, it is difficult to separate the lines. This results in a blended feature. However, the lines can be resolved with higher spectral resolution.

Certain chemical subgroups of atoms in a complex molecule, such as -CH_2-(methylene), -CH_3(methyl) etc., vibrate at characteristic stretching and bending frequencies for that group giving rise to absorption or emission lines at or near the same frequency regardless of the structure of the rest of the molecule. Therefore, infrared spectra can be used as a diagnostic of the chemical subgroup present in the material. The chemistry dominated by the cosmically abundant H,C,N and O atoms and the diagnostic bands from a combination of these bands fall almost exclusively in the near and mid-infrared region between 2.5 and $25\mu m$(4000 to 400 cm^{-1}). The wavelength coverage of near-infrared(NIR), mid-infrared (MIR) and far-infrared(FIR) are around 1-5μm, 5-30μm and 30-300μm, respectively. The various spectral signatures in the mid-infrared region arising out of subgroups composed of H,C,N and O is shown in Fig. 2.1. Figure 2.2 shows

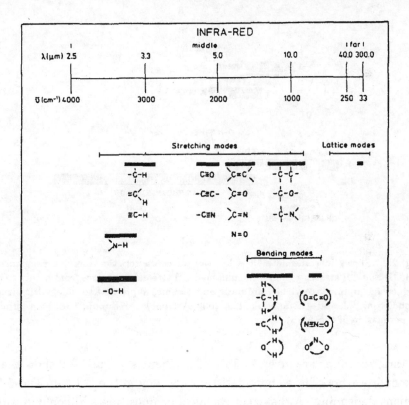

Fig. 2.1 The vibrational frequency ranges of various molecular groups (Allamandola, L.J. 1984, In *Galactic and Extragalactic Infrared Spectroscopy*, Eds. M.F. Kessler and J.P. Phillips, D. Reidel Publishing Company, p.5: with kind permission of Springer Science and Business Media).

in some detail the diagnostic spectral bands of subgroups comprising H,C,N and O in the wavelength region 2.8 - 3.8μm. The characteristic frequency

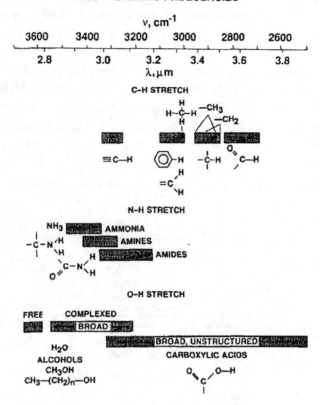

Fig. 2.2 Shows in some detail the location of characteristic vibration frequencies of NH, OH and CH stretch region. The unlabled CH stretching regions from higher to lower frequencies are for, acetylenic, aromatic and olefinic, aliphatic and aldehydric stretches (Pendleton, Y.J. and Allamandola, L.J. 2002, Astrophys. J. Suppl., **138**, 75: reproduced by permission of the AAS).

of various groups are given in Table 2.1. Figures 2.1 and 2.2 show that all the hydrocarbons have characteristic frequency in the 3-3.4μm and 5-8μm regions depending on the exact chemical composition and configuration of the larger molecules containing the spectrally active groups. Therefore, infrared spectra is used extensively to classify species and subgroups by chemical type, like aliphatic vs aromatic hydrocarbons etc. As for example, the 3.4μm emission feature seen from astronomial sources is generally attributed to C-H stretching modes of -CH_2- and -CH_3 groups in aliphatic hydrocarbons.

For estimating the column density of the absorber responsible for the

Table 2.1 Infrared integrated absorbance values for molecules of astrophysical interest in the solid state.

Molecule	Mode	λ (μm)	A (m/molecule)
H_2O	O-H stretch	3.05	2.0×10^{-18}
	H-O-H bend	6.0	8.4×10^{-20}
	libration	12	3.1×10^{-19}
NH_3	N-H stretch	2.96	2.2×10^{-19}
	deformation	6.16	4.7×10^{-20}
	inversion	9.35	1.7×10^{-19}
CH_4	C-H stretch	3.32	7.7×10^{-20}
	deformation	7.69	7.3×10^{-20}
CO	C=O stretch	4.67	1.1×10^{-19}
CO_2	C=O stretch	4.27	7.6×10^{-19}
	O=C=O bend	15.3	1.5×10^{-19}
CH_3OH	O-H stretch	3.08	1.3×10^{-18}
	C-H stretch	3.53	5.3×10^{-20}
	CH_3 deformation	6.85	1.2×10^{-19}
	CH_3 rock	8.85	1.8×10^{-20}
	C-O stretch	9.75	1.8×10^{-19}
H_2CO	C-H stretch (asym.)	3.47	2.7×10^{-20}
	C-H stretch (sym.)	3.54	3.7×10^{-20}
	C=O stretch	5.81	9.6×10^{-20}
	CH_2 scissor	6.69	3.9×10^{-20}
HCOOH	C=O stretch	5.85	6.7×10^{-19}
	CH deformation	7.25	2.6×10^{-20}
C_2H_6	C-H stretch	3.36	1.6×10^{-19}
	CH_3 deformation	6.85	6.0×10^{-20}
CH_3CN	C$\equiv$N stretch	4.41	3.0×10^{-20}
OCN^-	C$\equiv$N stretch	4.62	1.0×10^{-18}
H_2S	S-H stretch	3.93	2.9×10^{-19}
OCS	O=C=S stretch	4.93	1.5×10^{-18}
SO_2	S=O stretch	7.55	3.4×10^{-19}

Taken from Wittet, D.C.B. 2003, *Dust in the Galactic Environment*, Institute of Physics publishing, Bristol. For band strength A, Schutte, W.A. 1999, *Formation and Evolution of solids in space*, Eds. J.M. Greenberg and A. Li, Kluwer Academic Publishers, p.177: with kind permission of Springer Science and Business Media.

observed absorption band, a knowledge of the absorbance or band strength, A, is required. The incident radiation I_0 and the transmitted radiation I are related by

$$I = I_0 \exp(-\tau) \qquad (2.1)$$

where τ is the optical depth of the material between the source and the observer and is given by

$$\tau = clk. \qquad (2.2)$$

Here c is the concentration of the absorbers(molecules/cm^3), l(cm) is the path length of the infrared beam through the material and k is the absorption crosssection (cm^2). The absorption crosssection k, in general, is frequency dependent.

The intensity A of a vibrational band is defined as the integrated band strength or integrated absorbance per molecule

$$A = \frac{1}{cl} \int_{\nu_1}^{\nu_2} \ln\left(\frac{I_0}{I}\right) d\nu. \qquad (2.3)$$

Here ν is the frequency in cm^{-1} and the integration performed over the band. For a Lorentzian shaped band, the integral over the absorption band can be approximated as

$$\int_{\nu_1}^{\nu_2} \ln\left(\frac{I_0}{I}\right) d\nu \equiv \int \tau_\nu d\nu \approx \tau_{max} \Delta \nu_{1/2} \qquad (2.4)$$

for symmetrically shaped bands, where τ_{max} is $\ln(I_0/I)$ at the maximum of the band and $\Delta \nu_{1/2}$ is the full width of the band at half maximum(FWHM). Hence

$$A = \frac{\int \tau_\nu d\nu}{cl} = \frac{\int \tau_\nu d\nu}{N} \approx \frac{\tau_{max} \Delta \nu_{1/2}}{N}. \qquad (2.5)$$

Here $\tau(\nu)$ is the frequency dependent optical depth, τ_{max} is the maximum optical depth of the feature of interest. Hence the column density N(= cl) of the absorber responsible for a given absorption band can be calculated once A is known.

2.3 Mixed Ices

A large fraction of volatiles condense onto grains forming ice mantles in molecular clouds. Therefore, the mantle should contain many of the species seen in the gas phase. However, the mantle composition do not simply reflect the gas phase abundances due to surface reactions involving various species. They are also processed due to exposure to high energy radiation. Since H, C, N, and O are the most abundant elements in the universe, the molecules formed on the surface of grains involve mainly these species. Since hydrogen is more abundant than the heavier elements by 3 to 4 orders of magnitude, grain surface chemistry is largly dependent upon the local H/H_2 ratio. In regions where H/H_2 is large, the H atom addition(hydrogenation) dominates which leads to abundant molecules such as CH_4, NH_3 and H_2O. If on the other hand when H/H_2 is sustantially less than one, the atoms C, N and O can freely interact with each other giving rise to species such as CO, CO_2, O_2 and N_2. Therefore, two different types of mantle composition can be produced by surface grain reactions, one dominated by hydrogen-rich molecules and the other dominated by hydrogen-poor molecules. Irradiation and thermal processes will lead to more complex molecules. Figure 3.18 shows these possibilities in a schematic diagram. An example of the possible pathways involving CO and the simple photoproducts of an ice made up of H_2O, CH_4 and NH_3 is shown in Fig. 2.3.

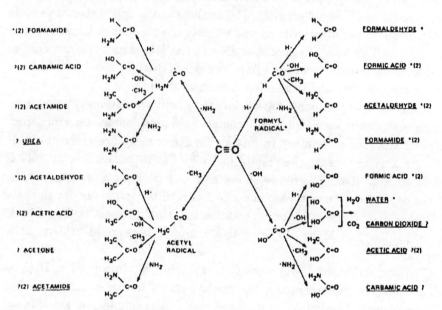

Fig. 2.3 Possible path ways involving CO in an interstellar grain mantle leading to complex molecules (Allamandola, L.J. and Sandford, S.A. 1988, In *Dust in the Universe*, Eds. M.E. Bailey and D.A. Williams, Cambridge University Press, Cambridge, p.229).

The general procedure used in the laboratory study of interstellar ices are as follows. The laboratory set up for the production of interstellar ice analogs involve a high vacuum system cooled to a temperature $\sim 10K$ by means of a cryostat. A flow of the gas mixture allows it to condense on a cold surface of aluminium mirror or sapphire window resulting in the formation of a layer of amorphous ice. The frozen gas mixture is subjected to different kinds of radiations depending upon the study under consideration. These radiations penetrate the sample and cause photochemical reactions in the bulk of the material and produce radicals and others. After this process is over the sample is allowed to warm up which makes the various components of the mixture to evaporate. The chemical processing of the released material are then carried out by several methods. They include among others, infrared absorbance spectrometry, mass spectrometry, ultraviolet-visible spectrometry etc. For irradiating the frozen gases, a hydrogen discharge ultraviolet lamp can be used to simulate photo-processing. Cosmic ray bombardment can be simulated by exposure to keV/MeV ions. This energy can easily be generated in small Van de Graaff accelerator. For stimulating grain surface chemistry between molecules and atoms, the frozen gases can be exposed to a flow of atomic hydrogen or oxygen by dissociating H_2 or O_2 in a discharge. The analysis of the spectral data provide information on the identification and number of new species formed during irradiation. The astrophysical spectra and the laboratory spectra carried out for relevant materials can then be compared directly inorder to identify the ice composition and determine molecular abundances.

A variety of experiments have been carried out on systems ranging from simple component samples to more complicated multi-component mixtures. Some of the studies relating to interstellar ice analogs are given in Table 2.2. As a typical case, the results for solid CO in mixed ices at 10K is shown in Fig. 2.4. The figure shows that the position and shape of the profile depend on the nature of the ice. The results carried out for the case of H_2O+CH_4 before and after irradiation to 1MeV protons at $T < 20K$ is shown in Fig. 2.5. Many of the synthesized molecules are identified in the spectra.

The formation of C_2H_6 and CH_4 in the experiment with $H_2O + CH_4$, helped in understanding the mechanism of formation of C_2H_6 and CH_4 in Comets Hyakutake and Hale-Bopp, where significant amounts of C_2H_6, CH_4 and C_2H_2 were detected. The ratio of $C_2H_6:CH_4$ seen in the H_2O+CH_4 experiments were smaller than those observed in comets. However, a higher $C_2H_6:CH_4$ ratio can be formed in the experiment with $H_2O+C_2H_2$ ices(5:1). Therefore the mechanism of formation of C_2H_6 in comets is consistent with laboratory findings in showing that it is through H atom addition reaction to C_2H_2. Firstly, H atoms come from the proton irradiation of H_2O and then add sequentially to C_2H_2 in steps producing C_2H_4 as the stable molecule and then C_2H_6.

The Infrared Space Observatory mid-infrared spectra of deeply embed-

Table 2.2 Infrared spectroscopy of interstellar ice analogs.

Reference	species	comment
Hagen et al. 1981	H_2O	pure
Hagen et al. 1983	numerous	mix with H_2O, low res.
d'Hendecourt & Allamandola '86	numerous	Pure ices, low resolution
Sandford & Allamandola 1988	CO	
Sandford & Allamandola 1990	CO_2	
Schutte et al. 1991	CH_3OH	only in H_2O, only CO str. mode
Ehrenfreund et al. 1992	O_2	
Hudgins et al. 1993	numerous	simple mixtures
Palumbo & Strazzulla 1993	CO	ion irr. ices
Palumbo et al. 1995	OCS	
Gerakines et al. 1995	CO, CO_2	Band strengths
Schutte et al. 1996a	CH_3OH, H_2CO	only CH stretch feat.
Ehrenfreund et al. 1996[a]	H_2O	in apolar ice mixtures
Bernstein et al. 1997	(iso-)nitriles	Numerous species, simple mix.
Boogert et al. 1997	CH_4, SO_2	
Ehrenfreund et al. 1997[a]	CO, CO_2	apolar ice mix.
Elsila et al. 1997	CO	apolar ice mixtures
Boudin et al. 1998	ethanol, ethane, H_2O_2	simple mixtures

Schutte, W.A. 1999, In *Laboratory Astrophysics and Space Research*, Eds. P. Ehrenfreund, K. Krafft, H. Kochan and V. Pirronello, Kulwer Academic Publishers, p.69: with kind permission of Springer Science and Business Media.

ded sources in molecular clouds has shown the presence of a large number of volatile solids (ices) (Fig. 3.15). Some of the condensed phase species identified in the interstellar ices are, H_2O, CO, CO_2, CH_3OH, H_2CO, CH_4, HCOOH, XCN, NH_3, N_2, O_2, O_3 etc. Therefore interstellar ices contain beside water ice, both the hydrogen-rich ices (called polar ices) such as CH_4, NH_3 and CH_3OH and hydrogen-poor ices (called apolar ices) such as O_2, N_2, CO and CO_2.

Several studies have been carried out on both polar and apolar ices. In the polar ice studies a mixture of molecules containing H_2O, CH_3OH, CO and NH_3, representative of interstellar grains, in various relative proportions were used (Fig. 2.6). After irradiating with ultraviolet photons, the new molecules CO, CO_2, CH_4, HCO, H_2CO etc. were found to be present. When the ices were warmed to a temperature of 200K, more molecules were produced. Some of them are, ethanol(CH_3CH_2OH), formamide (HC (=O) NH_2), acetamide (CH_3C (=O) NH_2) and molecules with C ≡ N groups (nitriles). When the residue was warmed to room temperature (300K), more species such as hexamethylenetetamine(HMT, $C_6H_{12}N_4$), ethers, alcohols

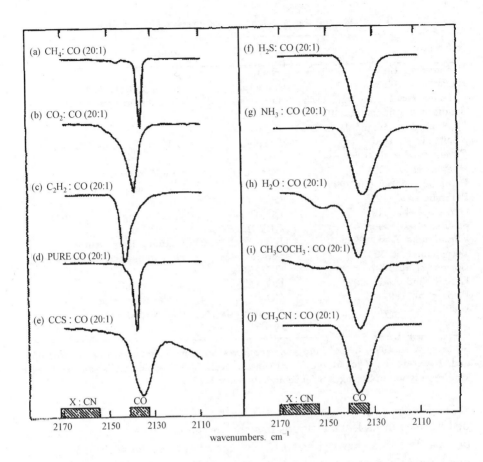

Fig. 2.4 The profile of solid CO in different ices (Sandford, S.A., Allamandola, L.J. Tielens, A.G.G.M. and Valero, G.J. 1988, Astrophys. J., **329**, 498: reproduced by permission of the AAS).

and compounds related to Polyoxymethylene [POM, $(-CH_2O-)_n$], ketones [R-C (=O)-R'] and amides [H_2NC (=O)-R] were produced. Most of the carbon in the residue is likely to have come from methanol in the original ices. Some studies on apolar ices such as (N_2 and CO), (N_2, O_2 and CO), (N_2, O_2, CO_2 and CO) and (N_2, H_2O, O and CO) etc have been carried out. When the samples were irradiated with UV photons, new molecules such as CO_2, N_2O, O_3, CO_3, HCO, H_2CO etc. were detected. The species produced in polar and apolar ices could be different when exposed to UV radiation, ion irradiation or for thermal process. It is likely that all the three processes are important in the astronomical environments one time or the other. Many of the species found in the photolysis residue are also

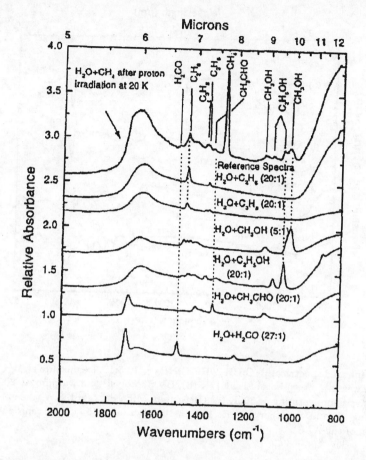

Fig. 2.5 Species observed after irradiation of H_2O + CH_4 at 20K(Moore, M.H. 1999, In *Solid Interstellar Matter: The ISO Revolution*, Eds. L. d'Hendecourt, C. Joblin and A. Jones, Springer-Verlag, p.199: with kind permission of Springer Science and Business Media).

present in comets. The presence of POM-like species in the photolysis residue supports the suggestion that the feature seen in the *in-situ* measurements of Comet Halley could be POM. Therefore, experiments have shown that complex molecules and organics can readily be formed in the ices from the interstellar medium. However, the exact composition of the ices in molecular clouds depend upon the local ratio, H/H_2. Therefore the species seen in the ice mantles from interstellar medium are due to chemical synthesis of complex molecules exposed to ultraviolet photons and galactic cosmic rays over a long period of time ($\sim 10^5 - 10^7$ yrs).

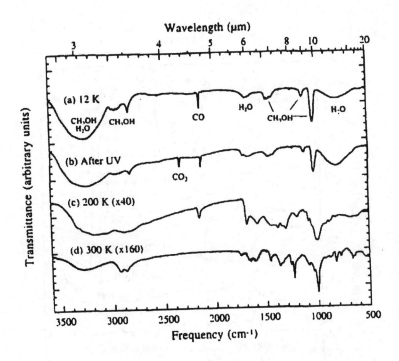

Fig. 2.6 Infrared spectra of $H_2O:CH_3OH:CO:NH_3 = 10:5:1:1$ ice mixture at 12K before UV photolysis (a), after photolysis at 12K (b), UV photolysis and warming to 200K (c), UV photolysis and warming to 300K (d) (Bernstein, M.P., Sandford, S.A., Allamandola, L.J., Chang, S. and Scharberg, M.A. 1995, Astrophys. J., **454**, 327: reproduced by permission of the AAS).

The results of the studies show that nature, position, shape, and profile of the spectral band depend upon whether the ice is crystalline or amorphous, icy mixture, percentage composition of species in the mixture, photon or ion irradiation with various dosages, cooling or warming of the samples etc. Extensive laboratory data is available for various mixtures relevant to interstellar and cometary conditions. Some studies on 'dirty ices' where refractory solids (minerals and silicates) are mixed with ices have also been carried out.

So far, a general qualitative discussion was given of the expected changes in the spectral features of icy mixtures. For more quantitative studies, the absorbance values or band strengths A, measured in various ice mixtures are required. For this purpose the infrared spectrum through a sample of known thickness and density is measured. A plot of $\ln(I_0(\nu)/I(\nu))$ versus frequency (cm^{-1}) is carried out, where $I_0(\nu)$ is the spectrum measured through the substrate before the sample deposition and $I(\nu)$ the spectrum measured after deposition. The mean value of the band strength or absorbance A can then be calculated. The integrated absorbance for the vibrational transitions of a number of astrophysically interesting materials are given in Table 2.1.

2.4 Silicate

2.4.1 *Structure of Silicate*

There are various types of silicates. They differ in their chemical composition and mineral structure. The basic building block of silicate material is SiO_4. They are tetrahedral in shape and has negative charge. In the tetrahedra, the four oxygen atoms of SiO_4 are located one at each of the corners, while Si atom is at its center. The two main forms of silicate material of astrophysical interest are the Olivines and Pyroxenes. The structure of silicates is shown in Fig. 2.7. Olivines can be considered as solid solutions of Mg_2SiO_4 and Fe_2SiO_4, while Pyroxenes of $MgSiO_3$ and $FeSiO_3$. They can be represented by a general formula as, $Mg_{2x}Fe_{2-2x}SiO_4$ for olivines and $Mg_xFe_{1-x}SiO_3$ for pyroxenes. The value of x can take values between 1 and 0. The end members of olivine class are Forsterite (x=1, Mg_2SiO_4) and Fayalite (x=0, Fe_2SiO_4). The corresponding ones for pyroxenes are Enstatite (x=1, $MgSiO_3$) and Ferrosilite (x=0, $FeSiO_3$). Two tetrahedra can be connected together by sharing an oxygen atom or connected by positively charged cations, like Mg^{2+}, Fe^{2+} etc. These two cases give different structural shapes: chain-like structure for pyroxene and island-like structure for olivine. In the case of amorphous silicate, long-range order is absent due to blending of structures and only short-range order exists. But crystalline silicate exhibit both short-range and long-range order. The degree of long-range order (amorphous to crystalline) has a considerable effect on the spectroscopic properties of the material. For example, amorphous silicates exhibit broad and smooth spectral features, while crystalline silicates show sharp and distinct features.

2.4.2 *Silicate Studies*

A wide variety of methods and techniques are available for producing silicate material in the laboratory. They include laser evaporation, evaporation with a mixture, evaporation from an electric arc, sol-gel reactions, plasma

Silicate structure

Pyroxenes: $MgFeSi_2O_6$ or
$Mg_xFe_{1-x}SiO_3$

Solid solution series of $MgSiO_3/FeSiO_3$:

$MgSiO_3$			$FeSiO_3$
Enstatite 0-5 Ma% FeO		Fe-Hypersthene	Ferrosilite
	Bronzite 5-15 Ma% FeO		<10 Ma%
	Hypersthene >15 Ma% FeO		MgO

Chain silicates: $[SiO_3]_\infty^{2-}$

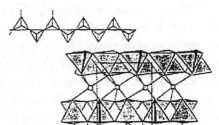

Olivines: $Mg_{2x}Fe_{2-2x}SiO_4$

Solid solution series of Mg_2SiO_4/Fe_2SiO_4:

Mg_2SiO_4			Fe_2SiO_4
Forsterite <10 Ma% FeO		Fe-Hortolonithe	Fayalite
	Olivine 10-30 Ma% FeO		<10 Ma%
	Hortolonithe		MgO

Neso silicates: $[SiO_4]^{4-}$

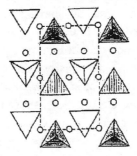

Fig. 2.7 Structure of Olivines and Pyroxenes (Henning, Th. 1999, In *Asymptotic Giant Branch Stars*, Eds. T. Le Bertre, A. Lebre and C. Waelkens, IAU Symposium 191, ASP Conference Publications, p.221:By the kind permission of the Astronomical Society of the Pacific Conference Series).

deposition etc. In the laser evaporation method, high power beam impinging on a target vaporizes the material, which is collected. The advantage of this method is that, it gives a homogeneous material. Vapour condensation is a common technique that is used extensively. The sol-gel method is a chemical process involving suspension of very small particles in a liquid. The particles grow through chemical reactions, forming solid aggregates. Scanning electron microscopy of the condensed grains show the structure of the particles to be of fluffy chain-like texture, formed mainly by grains with spherical shape. These resemble some of the structures seen from interplanetary dust particles.

The laboratory studies of silicates have emphasized on properties such as, physical, chemical, structural and optical properties and condensation etc. The effect of thermal annealing and exposure to UV radiation and high energy particles, which can modify the intrisic material properties are also studied. Several techniques are available for the study of the above properties. For qualitative and quantitative studies, Scanning Electron Microscope and the Energy Dispersive X-ray Spectrometer can be used. The information on shape, size and coagulation of nanometer size particles in smoke samples can be obtained from Transmission Electron Microscopy. X-ray Diffraction method can be used to study the structural properties of the material. Spectroscopic studies can be used effectively for the study of spectral behaviour of the materials.

2.4.2.1 *Condensation*

The condensation of gas to grains is a two step process. i.e. nucleation and growth. Firstly, multicomponent gaseous mixture has to first nucleate to produce seed nuclei, overwhich the material grows to give macroscopic solid particles. Since 1971, several controlled homogeneous experiments have been carried out in several laboratories on refractory materials. They are summarized in Table 2.3. The experiments cover a wide range of temperature and pressure. They have been carried out on several systems such as, (i) simple metals Na, Fe, Si, Yb, Pb and Bi (ii) binary mixtures Fe/Si and (iii) oxides, FeO_x, SiO_x, Mg_xSiO_y and Fe_xSiO_4.

In the vapour phase nucleation experiments carried out on Mg-SiO(Mg+SiO) systems, a metastable condensate seems to form, either Si_2O_3 or amorphous magnesium silicate. This had been noticed in the earlier experiments as well. The results also showed that for the Mg-SiO system, the critical partial pressure of SiO needed to initiate nucleation

Table 2.3 Nucleation experiments carried out by various groups on refractory materials.

Condensate	Temperature Range (K)	Method: Reactant in Background Gas	Reference
Fe···	$1800 < T < 2150$	Shock tube:$Fe(CO)_5$ in Ar	Kung & Bauer '71
	$1600 < T < 1800$	Shock tube:$Fe(CO)_5$ in Ar	Freund & Bauer '77
	$1625 < T < 2125$	Shock tube:$Fe(CO)_5$ in Ar	Frurip & Bauer '77
	$1900 < T < 2400$	Shock tube:$Fe(CO)_5$ in Ar	Stephens & Bauer '81
	$1000 < T < 1700$	Shock tube:$Fe(CO)_5$ in Ar	Steinwandel et al. '81
Pb···	$950 < T < 1225$	Shock tube:$Pb(CH_3)_4$ in Ar	Frurip & Bauer '77
Bi···	$755 < T < 1275$	Shock tube:$Bi(CH_3)_3$ in Ar	Frurip & Bauer '77
Si···	$1500 < T < 2800$	Shock tube:SiH_4 in Ar	Tabayashi & Bauer '79
	$1550 < T < 2400$	Shock tube:SiH_4 in Ar	Stephens & Bauer '81
	$1500 < T < 3000$	Shock tube:SiH_4 in Ar	Steinwandel et al. '82
Yb···	$500 < T < 625$	Gas evaporation:Yb in Ar	Onaka & Arnold '81
Na···	$350 < T < 400$	Gas evaporation:Na in Ar	Hecht '79
Fe/Si···	$1650 < T < 2400$	Shock tube:$Fe(CO)_5$ +SiH_4 in Ar	Stephens & Bauer '81
FeO_x···	$1950 < T < 2900$	Shock tube:$Fe(CO)_5$ +N_2O in Ar	Stephens & Bauer '81
SiO_x···	$1250 < T < 4200$	Shock tube:SiH_4 +N_2O (+H_2) in Ar	Stephens & Bauer '81
	$750 < T < 1015$	Gas evaporation:SiO in H_2	Nuth & Donn '82
Mg_xSiO_y···	$750 < T < 1015$	Gas evaporation:SiO in H_2 +Mg	Nuth & Donn '83
Fe_xSiO_y···	$1600 < T < 4000$	Shock tube:$Fe(CO)_5$+SiH_4 +N_2O in Ar+CO_2+H_2	Stephens & Bauer '83

Donn, B.D. and Nuth, J.A. 1985, Astrophys. J., **288**, 187: reproduced by permission of the AAS.

was the same as the pure SiO system for temperature above 925K. But below 900K, significantly lower pressure of SiO was required to initiate nucleation. These results indicate that for $T < 925K$, the nucleation was triggered by the formation of mixed Mg-SiO clusters, while for $T > 950K$, the nucleation is controlled by pure $(SiO)_x$ clusters. Similar results were also obtained from the shock tube experiment, which contained a mixture of Fe, Si, C, O, N and H where the temperature ranged from 1600 to 4000K.

All these experiments have a great bearing on the application of classical nucleation theory for the growth of grains in astrophysical situations (Sec. 2.9).

2.4.2.2 Effect of Environment

Amorphous silicate is the most common form present in astrophysical objects. Infact, the dominant component of silicate in the interstellar medium is amorphous silicate. Therefore, the most unexpected result that came out of observations carried out with ISO, is the presence of crystalline silicate in most of the astrophysical objects, including comets. Therefore, it is important to understand how amorphous silicate could be transferred to crystalline silicate.

The transformation of amorphous to crystalline silicate is possible by the process of annealing, in which the ordered arrangement of silicate tetrahedra is brought about by atomic diffusion. In this process, thermal diffusion brings about rearrangement of the structural units (tetrahedra) which results in long-range order. Diffusion in solids is a result of activation of lattice defects. Therefore, the poorly ordered state of the amorphous material slowly drifts towards more energetically favourable positions leading to macro-crystalline structure.

Several experiments have been carried out to see the actual change over from the amorphous state to crystalline state. Magnesium silicate smokes inside a vacuum was monitored for structural changes through infrared spectroscopy, as the temperature varied between 1000 and 1048K. The observations showed a broad feature at 9.3μm in the starting material. After annealing at 1027K, the spectra showed two features at 9.8 and 11.0μm. These features and the spectral signature at 20μm, showed the formation of olivine.

Annealing experiments carried out on magnesium, magnesium-iron and iron silicates showed that iron silicates were found to produce silica and fayalite, while magnesium silicates produced silica, enstatite and forsterite.

Annealing of pure magnesium amorphous smoke can lead to complete crystallization in a few hours, while $MgSiO_3$ required a day for change over to crystalline state. As an example, Fig. 2.8 shows the results of annealing experiment on diopside ($CaMgSi_2O_6$) smoke. As can be seen from the figure, the time required to change over from amorphous to crystalline diopside is less than one hour at T=1050K.

All the experiments have shown that annealing of the silicate material, can easily transform the material from the amorphous state to crystalline state in a relatively short period of time, $\sim$ several hours at T=1000K. The structural evolution of amorphous magnesium and iron silicates during the annealing process, has been studied. The results show that thermal annealing does not alter the bond length or angles within the SiO_4 tetrahedra, but simply leads to ordered arrangement of the SiO_4 units.

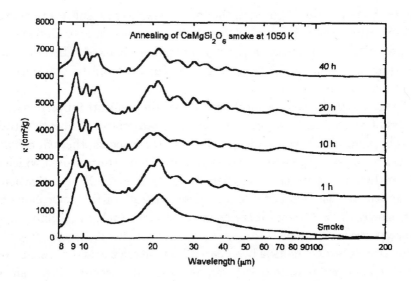

Fig. 2.8 Annealing of $CaMgSi_2O_6$ smoke at 1050K and for times from 1 to 40 hours. The curves for 1 to 40 hours are shifted vertically for clarity (Colangeli, L., Henning, Th., Brucato, J.R. et al. 2003, Astron. Astrophys. Rev., 11, 97).

2.4.2.3 Silicon Nano-crystals

Solids give photoluminescence in the visible region, when exposed to UV radiation. This is a common property of solids. However, bulk material is known to fluoresce very weakly. Even though bulk crystalline Si is known to fluoresce weakly, small size porous grains (nanometer size) produce strong photoluminescence spectrum in the visible region at room temperature. This was first noticed in 1990 from porous silicon, which was made by electrochemical etching of silicon wafers. The photoluminescence from Si nano-crystals have been studied in the laboratory for different size particles. The photoluminescence emission feature is found to systematically shift to bluer region with increase in the size of the particles. For example, the variation in size of Si nano-crystals between 2 and 9nm, the peak of the photo luminescence feature varies from 5060 to 9200Å. Temperature has some effect on the peak of the photoluminescence spectrum. With a decrease in temperature, the peak shifts to the bluer side. A decrease of temperature from 500 to 100K, induces roughly a shift of the peak of around 70nm. Crystalline Si in the form of nano-clusters in the size range 3 to 5nm can give the peak of emission at 7000Å, corresponding to Extended Red Emission seen from astrophysical objects (Sec. 3.7.3).

The mechanism of high efficiency for photoluminescence from Si nanocrystals is reasonably understood from theoretical means. Nanometer size porous silicon can be considered as random network of interconnected silicon crystals. A carrier inside the nano-size volume, is not affected by the entire volume, but instead considered being inside a potential well. Therefore, its energy level depend strongly on the size of the particle. Hence, the energy of the emitted photoluminescence becomes a function of the particle size of the particle in such a way that the peak emission shifts bluewards as the size of the particle decreases. In such a small volume, once the UV photon is absorbed and excitation takes place, the probability of recombining via a radiation transition is very high. With the result the photoluminescence efficiency reaches almost 100%.

In view of the importance of nano-sized particles in the study of dust in astrophysical environments, the synthesis and characterization of nano-sized particles has assumed great importance.

2.4.2.4 *Optical Constants*

The refractive indices ($\mu=n+ik$) of silicate materials are essential for the study of scattering properties from grains. They have been determined for various kinds of silicate materials. The results are also available for various Mg/Fe ratios (Sec. 2.7.2).

Similarly, the mass absorption coefficients required for quantitative studies have also been determined for various types of silicates. They are also available for different Mg/Fe ratios (Sec. 2.7.3).

2.5 Carbon

2.5.1 *Forms of Carbon*

Carbon exists in different forms, varying from carbon atoms to hydrocarbons(about 50 to few hundred carbon atoms) and finally to solids. Carbon can exist both in crystalline and amorphous forms. Some of the forms of carbon are, graphite, diamond, coal, soot, amorphous carbon, hydrogenated amorphous carbon (HAC), quenched-carbonaceous condensates (QCC), polycyclic aromatic hydrocarbons (PAHs) and carbon clusters. The large variety and the differing properties of carbonaceous materials is due to its ability to bond adjacent atoms in three different hybridization configurations, sp, sp^2 and sp^3 (via single, double and triple bonds). Here s and

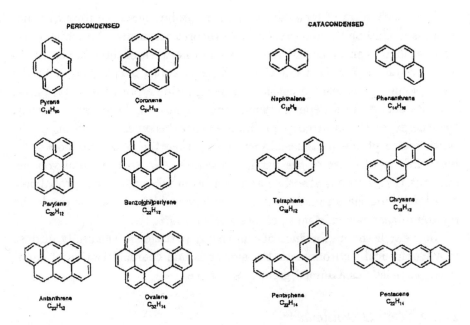

Fig. 2.9 Molecular structures of some representative PAHs (Salama, F. 1999, In *Solid Interstellar Matter: The ISO Revolution*, Eds. L. d'Hendcourt, C. Joblin, and A. Jones, Springer-Verlag, p.65: with kind permission of Springer Science and Business Media).

p refer to orbital electrons. Aliphatic molecules have open structure, while Aromatic molecules have ring-like structure, like Benzene, C_2H_6. Benzene is the building block for a wide range of hydrocarbon molecules. Infact, the basis of almost all of the work on unusual carbon molecules is the carbon ring. Molecules with several aromatic rings of sp^2-hybridized carbon atoms are called Polycyclic Aromatic Hydrocarbons and abbreviated as PAH (Fig. 2.9). PAHs are the most thermodynamically stable hydrocarbon compounds that exist in the gaseous form. The solid carbon can exist as graphite, diamond and amorphous carbon. Graphite is composed of sp^2 carbon atoms comprising PAH layers and the PAH layers are held parallel to each other by van der Waals forces. Diamond is formed out of sp^3 carbon atoms arranged in a manner referred to as a diamond lattice. A measure of the degree of disorder of the carbon structure is given by the sp^2/sp^3 ratio and the hydrogen content. On the other hand amorphous carbon does not exhibit a long range order in their arrangement. There are several classes of materials in this category which depend mostly on their appearance, for example, glassy carbon or soot. Others are characterized by their hydrogen content. For example, Diamond Like Carbon (DLC) is associated with low hydrogen content while Hydrogenated amorphous carbon (HAC) has larger hydrogen content. Hydrogenation means that a multiring structure such as PAH has a portion, which is not aromatic as double bonds are absent and instead two hydrogen atoms are attached to peripherical carbons. The formation of hydrogenation is an easy process as several laboratory studies and computer modeling have shown. Two important changes occur when a complex PAH structure becomes hydrogenated. They are, conversion of planar to nonplanar geometry and the loss of aromaticity accompained by loss of $\pi \rightarrow \pi^*$ electronic transitions. Kerogens are another class of amorphous carbonaceous materials with macromolecular structure composed of PAHs with aliphatic links. The transition from gas to solid involves the formation of clusters. One class of such clusters is the fullerenes (C_{60}) or nanotubes. Fullerenes are five-membered rings which give rise to curvature and is distinguished from PAHs which are planar in shapes. In terms of stability, planar PAHs are more stable than fullerenes as their curvature strains their structures which make them less stable.

PAHs are broadly grouped into two classes called compact (Pericondensed) and non-compact (Catacondensed) (Fig. 2.9). In the compact PAHs, some carbon atoms belong to three rings and in non-compact PAHs, no carbon atom belongs to more than two rings. PAHs can also exist in neutral and ionized states. The size of PAHs could be around 25 to 100 carbon atoms. Therefore wide varities of PAHs can exist under astrophysical conditions

2.5.2 Carbon Studies

Several groups are actively involved in laboratory experiments aimed at the study of formation and evolution of carbon dust under astrophysical conditions. This involves the study of formation of various types of carbon under different physical conditions, their chemical, structural and optical properties, effect on UV radiation and ion processing. Extensive studies have been carried out on all these aspects.

Carbon grains of submicron size can be produced by various methods. It can be produced in arc discharge between carbon/graphite electrodes in an inert argon atmosphere, burning of hydrocarbons such as benzene etc. The grains formed in condensation experiments usually occur in aggregated forms in contrast to carbon grains in astrophysical environments where it could occur as isolated grains. Therefore the optical behaviour of the condensing grains in the laboratory has to be properly used when applied to grains under astrophysical environment.

Several methods have been used for the study of structure, morphology, chemistry and optical properties of carbon grains. The internal structure can be studied using high-resolution transmission electron microscopy, X-ray diffraction and Raman spectroscopy. Among these, transmission electron microscopy is the best method. Electron energy loss, nuclear magnetic resonance and optical spectroscopy (from the ultraviolet to infrared) can be used to characterize the sp^2/sp^3 ratio, hydrogen content and the band structure.

Direct electronic interband transitions between the valency and conduction bands give rise to photon absorption. The strength, positions and profile of the absorption features are defined by the density of states. The $\sigma \rightarrow \sigma^*$ electronic transitions give rise to a feature located in the far ultraviolet region of 0.08-0.09μm, whereas the feature arising out of $\pi \rightarrow \pi^*$ electronic transition lie in the range of 0.20-0.26μm. The feature present in the interstellar extinction curve at $\lambda = 0.2175\mu$m is generally attributed to the feature arising out of $\pi \rightarrow \pi^*$ electronic transitions. Experiments have also been carried out to see the modification induced by some of the physical processes. Thermal annealing in vacuum of hydrogenated carbon grains produces modifications in the infrared spectral properties compatible with a progressive aromatization of the carbon structure. In particular, the structure of the grains changes from a mainly aliphatic to a dominant aromatic status. i.e. the carbon network tends to arrange in an increasing number of larger sp^2 clusters. The UV irradiated hydrogenated carbon produce effects similar to the thermal annealing. The feature at 0.2175μm is also present. Ion bombardment of hydrogenated carbon also induces 0.2175μm feature.

Nano-diamond particles can be formed by vapour deposition in the C-rich ejecta in stellar environment with high velocity grain-grain collisions. It may also form in supernova Type II. In the experiments carried out in the laboratory, synthetic nano-diamond particles of sizes $\sim 0.1\mu$m in diameter, were irradiated with a flux of atomic hydrogen at room temperature. In this process, the hydrogen atom attaches to carbon atoms forming CH bonds. The infrared absorption spectra of the irradiated sample showed the presence of two groups of features peaking at 3.43 and 3.53μm (Fig. 6.6). These features are attributed to CH stretching of H-terminated diamond. These results are of interest, as features at these wavelengths have been seen in the infrared spectra of Herbig Ae/Be stars (Sec. 6.5.5.1).

An experimental study was carried out on nanometer size soot particles(pure and hydrogenated carbon) that were frozen in a noble gas matrix in order to obtain spectra of the isolated grains. The laboratory spectra could satisfactorily reproduce the RCrB(R Coronae Borealis) type UV band observed at 0.24-0.25μm as well as the interstellar feature at 2175Å.

Carbon rich stars show strong emission feature at 11.3 μm. This feature has been attributed to SiC particles present in the circumstellar shells of these stars. The observed 11.3 μm feature show some variability which may offer the possibility to characterise more closely the composition and structure of SiC dust grains. With this in view, experimental studies in the infrared region have been carried out on a large number of samples of SiC particles. The laboratory spectra show a wide range of SiC features in the 10-13 μm region, both in peak wavelength and band shape. These are related to grain size, shape and to impurities in the material. The band profile is also found to be not dependent on the crystal type(α -vs β-SiC).

2.6 Polycyclic Aromatic Hydrocarbons (PAHs)

2.6.1 *Spectra*

As mentioned earlier, Polycyclic Aromatic Hydrocarbon molecules contain several aromatic rings (Benzene type, C_2H_6). PAHs can exist in various forms, such as neutral, ionized, compact, noncompact, hydrogenated, dehydrogenated etc. The neutral and ionized PAHs have been invoked to explain the spectral features seen in absorption and in emission from astronomical sources. This is due to the fact that PAHs possess, rich spectra with specific spectral signatures, a high photostability and also contain cosmically abundant carbon. Several experimental techniques are available for the study of PAHs, like suspension of the material of interest in solid and liquid environment, the isolation of the material in condensed inert-gas matrices (Matrix Isolation Spectroscopy, MIS) etc. Among all the techniquies, MIS has been used extensively for the study of neutral and ionized PAHs. In this method the PAH molecules are kept at low temperature (5-10K), isolated from each other and are trapped in a solid case. Neon is generally used as the matrix because of its low polarizability.

There existed already the absorption spectra of a large number of neutral PAHs ($\leq$25 carbon atoms) in the visible and UV region, even before their importance in astrophysics was known. They showed a series of discrete absorption bands in the UV and near UV range with relatively small absorption in the visible range. A typical spectra of neutral PAH is shown in Fig. 2.10. These absorption bands are associated with vibronic transi-

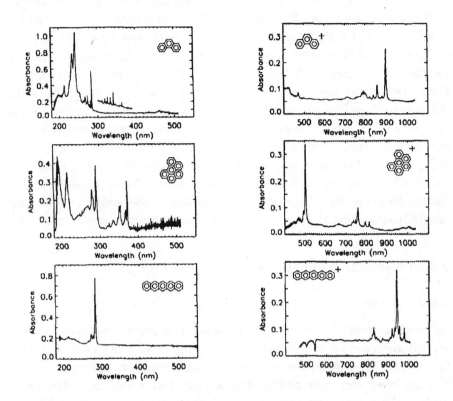

Fig. 2.10 (Left) Absorption spectra of neutral PAHs isolated in a neon matrix at 4.2K in the UV-Visible region. Top ($C_{14}H_{10}$), middle ($C_{22}H_{12}$,) bottom ($C_{22}H_{14}$) (Salama, F. 1999, In *Solid Interstellar Matter: The ISO Revolution*, Eds. L. d'Hendecourt, C. Joblin and A. Jones, p.65: with kind permission of Springer Science and Business Media).
(Right) Absorption spectra of PAH cations isolated in a neon matrix at 4.2K in the visible-near infrared region. Top ($C_{14}H_{10}^+$), middle ($C_{22}H_{12}^+$), bottom ($C_{22}H_{14}^+$) (Salama, F. 1999, In *Solid Interstellar Matter: ISO Revolution*, Eds. L. d'Hendecourt, C. Joblin and A. Jones, p.65: with kind permission of Springer Science and Business Media).

tions between the electronic states of the molecule. These studies have been extended enormously in recent times. A substantial amount of information has come out on the absorption spectra for neutral PAHs. For example, the absorption band energies of neutral PAHs shift towards longer wavelengths when the molecular size increases. Also for the case of naphthalene, the gas phase-to-matrix relative shift in energy is 0.25% in neon, while it is 2.5% shift, for more polarizable argon matrix. Irradiation of neutral PAHs with UV radiation produce new spectral features in the UV and near infrared range, as can be seen from Fig. 2.10. All the new features are found to be associated with PAH cation (PAH$^+$) formed by direct one-photon ionization of the neutral precursor. The perturbation induced by the solid matrix is more important for ions compared to neutral molecules. For example, the strongest absorption of pyrene cation $C_{16}H_{10}^+$ which is at 0.4395μm in a Ne matrix shifts to 0.4435μm in argon matrix and to 0.4500μm in an organic matrix.

The mid infrared region (2 to 25 micron) is where the fundamental vibrational modes of PAHs, their overtones are active (Table 2.1). The emission spectra of the gas phase of PAHs are becoming available. Studies have been extended to the spectra of ionized PAHs. The difference in the infrared spectra between the neutral and the ionized PAHs in the 2-16μm region is shown in Fig. 2.11. In this figure, the result of the addition of six PAH absorption spectra of neutrals (top) and the spectra for the corresponding ions (bottom) is shown. The figure shows that for neutral PAHs, the C-H out-of-plane bending features in the region of 11μm is quite strong. While the features in the aromatic C-C stretching region around 6μm and in the C-H in-plane bending region around 8.7μm are very weak. But for the spectra of ions, it is opposite to that of neutrals in that, the C-C stretch and C-H in-plane bend dominates.

The infrared spectra in the wavelength region 5-15μm of various kinds of galactic objects and galaxies can be modeled satisfactorily based on the combined laboratory spectra of neutrals and/or positively charged PAHs. For example, good match is obtained for the proto-planetary nebula IRAS 22272+5435, with a combination of 60% neutrals and 40% ionized PAHs. IRAS 22272+5435 is a C-rich object evolving from AGB phase to planetary nebula phase. Similarly, the emission spectrum from the Orion ionization ridge require a combination of completely ionized PAHs. As an illustration, comparison between the observed emission spectrum of the compact blue dwarf galaxy Haro 3 with a composite spectrum produced from an ionized PAH mixture is shown in Fig. 2.12. Haro 3 belongs to a class of Blue compact dwarf galaxies, which are small in size($\leq$5kpc diameter) and have ultraviolet excess. Haro 3 contain several starburst regions, dominated by Wolf-Rayet stars.

The studies of PAHs in the far infrared region are still limited. The FIR region (20-1000μm) is associated with the bending motions (out-of-plane deformation) of the PAH molecules. A study of the superposed absorption spectra of $\sim$ 40 PAH molecules in the wavelength region 14-40μm has been carried out. The absorption and emission spectra of a limited number of

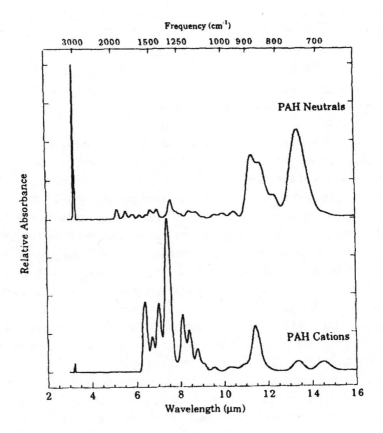

Fig. 2.11 Resultant absorption spectra of addition of 6 PAHs in neutral and ionic form to show the effect between the two. PAHs included are $C_{14}H_{10}$, $C_{16}H_{10}$, $C_{18}H_{12}$ (tetracene), $C_{18}H_{12}$ (1-2-Benzanthracene), $C_{18}H_{12}$ (crysene) and $C_{24}H_{12}$ (Allamandola, L.J., Hudgins, D.M. and Sandford, S.A. 1999, Astrophys. J., **511**, L115: reproduced by permission of the AAS).

PAHs has also been measured in the gas phase between 2.5 to 200 μm. The results of these studies indicate the presence of four new features at 16.2, 18.2, 21.2 and 23.1 μm. The mechanism of formation of these bands in the infrared region is through infrared fluorescence. When the molecule absorbs a UV photon, it leaves the molecule in highly excited states. The excited molecule will then relax by emitting infrared photons at its fundamental vibration frequencies (infrared fluorescence). The relative band strengths depend on both the size of the molecule and its vibrational energy content.

Due to the concentrated laboratory effort of several experimental

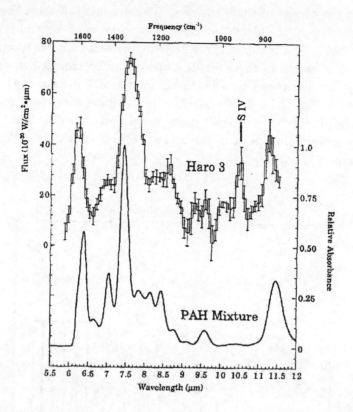

Fig. 2.12 Comparison of the emission spectrum from the compact blue dwarf galaxy Haro 3 with the composite absorption spectrum of a mixture of ionized PAHs. The mixture consists of 19% dicoronylene$^+$, 14% coronene$^+$, 11% anthracene$^+$, 9% benzo(k) fluoranthene$^+$, 9% chrysene$^+$, 9% phenanthrene$^+$, 9% pyrene$^+$, 6% 9, 10-dihydrobenzo(e)pyrene$^+$, 4% pentacene$^+$, 4% tetracene$^+$, 2% 1, 2-benzanthracene$^+$, 2% benzo(e)pyrene$^+$ and 2% naphthalene$^+$ (Salama, F. 1999, In *Solid Interstellar Matter: ISO Revolution*, Eds. L. d'Hendecourt, C. Joblin and A. Jones, p.65: with kind permission of Springer Science and Business Media).

groups, there exists lot of information on the structural properties of neutral and ionized PAHs containing from 10 carbon atoms upto about 50 carbon atoms over a wide range of wavelengths.

A series of emission lines at 3.3, 6.2, 7.7, 8.6 and 11.3 μm was first detected in 1970's in the spectra of HII regions and planetary nebulae. Since it could not be identified, it was called as Unidentified Infrared Features (UIR). Since then, these features have been seen from various objects in the Galaxy and from extragalactic sources. This indicates the common nature of the carrier of these features. The detection of these features in such diverse type of objects indicated that they must be widespread as well as stable. The features are generic in the sense that they appear as a group. The carrier of these features appears to be aromatic in character and they are attributed to vibrational modes of C-H and C-C. This is illustrated in Fig. 2.13. It is suggestive to attribute these features arising from PAHs. However, no specific molecule has been identified and also the emission arises from a host of molecules. Therefore, the identification of these features is not clear.

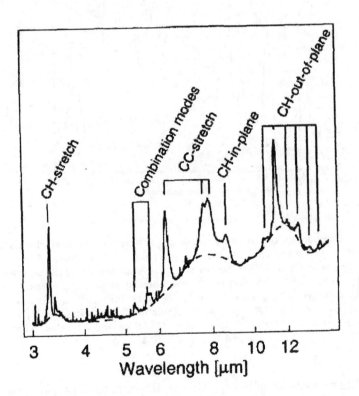

Fig. 2.13 Shows the assignment of the observed AUIB features in the planetary nebula NGC 7027 to vibration modes of aromatic molecules (Bakes. E.L.O., Tielens, A.G.G.M. and Bauschlicher, C.W. 2001, Astrophys. J., **556**, 501: reproduced by permission of the AAS).

2.6.2 Mass Spectra

Several Mass spectrometry studies have been carried out on a few selected PAHs for the study of chemical reactivity, photofragmentation and formation mechanisms. The electron attachment studies indicate low rates indicating the importance of PAH ions in cloud chemistry. The PAH ion-atom reaction rates indicate the rapid formation of protonated PAHs. The fragmentation mechanisms of small PAHs like anthracene ($C_{14}H_{10}$) and pyrene ($C_{16}H_{10}$) have been studied. They indicate that the dominant fragmentation channel for pyrene is dehydrogenation while for anthracene it is through the loss of C_2H_2. The same experiment also indicated that dehydrogenated PAHs appear to be more stable. This implies that compact PAHs may be more dominant than non-compact species in the interstellar medium.

The laboratory studies of PAHs have improved substantially in recent years. But more needs to be done. In particular the physical and chemical properties and UV and IR spectroscopy of PAHs containing around 50 to 100 carbon atoms are needed. These are difficult studies as it is harder to produce such large PAHs in the laboratory. These are also required for the study of size distribution of interstellar PAHs. The above studies relating to hydrogenated and dehydrogenated PAHs are also of interest.

Theoretical studies of PAHs are also required for supplementing experimental studies. These relate to, stabilty, structure, role of the groups, effect of hydrogenation, photodissociation rates etc. Several theoretical studies have been carried out for the study of PAH properties as a function of their molecular size (number of carbon atoms), their structure, their ionization state (positively or negatively charged ions) and the degree of hydrogenation. The results of the calculations indicate that PAHs with <50 carbon atoms can be destroyed or fragmented in UV dominated regions. The dominant channel for photodistruction is the loss of carbon through the photoejection of an acetylene (C_2H_2) molecule.

2.6.3 Steller Environment

The first molecules that are expected to be formed in stellar wind are naphthalene ($C_{10}H_8$), acenaphthylene ($C_{12}H_8$), acenaphthene ($C_{12}H_{10}$) and pyrene ($C_{16}H_{10}$). They are present in energetic environment. For simulating such conditions, naphthalene was coated on the inner walls of a sapphire tube and then exposed to plasma environment (discharge tube) containing

H, He, N or O. The resulting product that reside in the sapphire tube has a yellow colour and hence usually termed yellow residue. The resulting products were then subjected to chemical analysis. These showed the richness of the complex species which are present. These were beyond the products expected under thermodynamic equilibrium models. Substantial amount of hydrogenated PAHs which are described as hybrids of polycyclic aromatic (sp^2) and polycyclic aliphatic (sp^3) species were produced. Ultraviolet absorption spectroscopy of the plasma products, exhibit a broad feature centred at 2175Å, similar to the feature seen in the interstellar extinction curve (Fig. 2.14). The infrared absorption spectroscopy in the

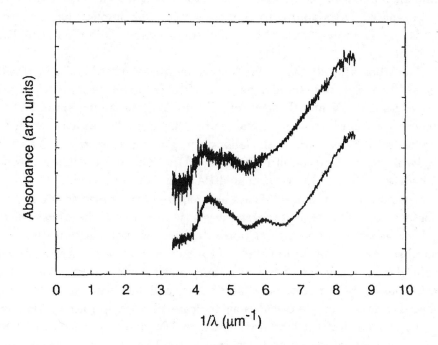

Fig. 2.14 The spectral feature produced around 2200Å when naphthalene is irradiated in a plasma reactor (Arnoult, K.M., Wdowiak, T.J. and Beegle, L.W. 2000, Astrophys. J., **535**, 815: reproduced by permission of the AAS).

C-H region (3.3μm, 3000 cm^{-1}) of the reaction yellow residue also show a feature which is similar to the feature seen from astronomical objects. As an illustration, a comparison of the laboratory spectra with the observed spectra in the proto-planetary nebula IRAS 05341+0852 is shown in Fig. 2.15. Mass spectroscopy of the residue shows the complete array of PAH species produced with molecular weights as high as 574. Some of the as-

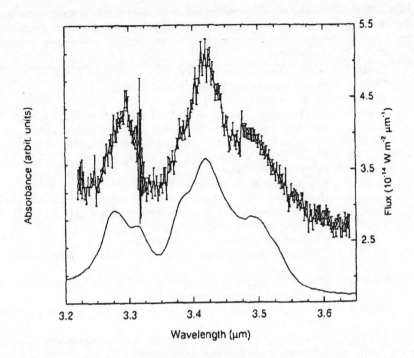

Fig. 2.15 Infrared spectra of the C-H stretch region of a film on KBr prepared from yellow residue formed from the sapphire tube (lower) and the emission spectra of protoplanetary nebula IRAS 05341+0852 (upper). The bands at 3.4µm can be attributed to alkane and at 3.3µm to aromatic molecular species (Beegle, L.W., Wdowiak, T.J., Robinson, M.S., Cronin, J.R., McGehee, M.D., Clemett, S.J. and Gillette, S. 1997, Astrophys. J., **487**, 976: reproduced by permission of the AAS).

signments are shown in Table 2.4.

The 3.4µm feature, attributed to the aliphatic C-H stretch vibration, is found to depend upon the environment, both with regard to the gaseous material and the radiation field present. Therefore, several experimental studies have been carried out under different conditions. For simulating cosmic ray processes in the interstellar medium, hydrogenated carbon atoms with different hydrogen content were irradiated with 30keV He^+. The results indicated an evolutionary transformation of the aliphatic component carrier by grain processing, leading to changes in the 3.4µm absorption feature. This result is consistent with the observation that the 3.4µm absorption band is present in the diffuse medium but is absent in the dense cloud environment. This difference could arise due to the fact that while destruction of the carrier component of the band can take place in both dense

Table 2.4 Mass distribution of a film evaporated onto quartz prepared from yellow residue obtained from sapphire tube in which hydrogen was used to create the plasma experiment.

Mass	Formula	Assignment
128···	$C_{10}H_8$	Naphthalene (aromatic double ring)
142···	$C_{11}H_{10}$	Methylnaphthalene (aromatic double ring)
152···	$C_{12}H_8$	Acenaphthylene
154···	$C_{12}H_{10}$	Acenaphthene (hydrogenated PAH with aromatic double ring)
156···	$C_{12}H_{12}$	Dimethylnaphthalene, ethylnaphthalene (aromatic double ring)
168···	$C_{13}H_{12}$	Phenalan (hydrogenated PAH with aromatic double ring)
178···	$C_{14}H_{10}$	Phenanthrene, anthracene
202···	$C_{16}H_{10}$	Pyrene, fluoranthene
204···	$C_{16}H_{12}$	Phenylnaphthalene (aromatic double ring)
208···	$C_{16}H_{16}$	Hexahydropyrene (hydrogenated PAH with aromatic double ring)
218···	$C_{17}H_{14}$	Phenyl, methylnaphthalene (aromatic double ring)
228···	$C_{18}H_{12}$	Chrysene, naphthacene, triphenylene
252···	$C_{20}H_{12}$	Perylene, benzopyrene, benzofluoranthene
254···	$C_{20}H_{14}$	Binaphthyl (multiaromatic double ring, GCMS indicates three isomers present)
256···	$C_{20}H_{16}$	Dihydrobinaphthyl (hydrogenated PAH with aromatic double ring, GCMS indicates three isomers present)
258···	$C_{20}H_{18}$	Tetrahydrobinaphthyl (hydrogenated PAH with aromatic double ring)
268···	$C_{21}H_{16}$	Methylbinaphthyl, dinaphthylmethane (multiaromatic double ring)
278···	$C_{22}H_{14}$	Dibenzo(a,h) anthracene, benzoperylene
282···	$C_{22}H_{18}$	Dimethylbinaphthyl, dinaphthylethane (multiaromatic double ring)
378···	$C_{30}H_{18}$	Naphthyl substituents on 252 (multiaromatic double ring)
380···	$C_{30}H_{20}$	Trinaphthyl (multiaromatic double ring)
434···	$C_{34}H_{26}$	Alkyl naphthyl-substituents on 252

Beegle, L.W., Wdowiak, T.J., Robinson, M.S. et al. 1997, Astrophys. J., **487**, 976: reproduced by permission of the AAS.

and diffuse regions due to cosmic rays and UV radiation, the formation of rehydrogenation of carbon atoms is harder in denser regions.

When Hydrogenated Amorphous Carbon (HAC) was produced in the presence of non-hydrocarbon gases, such as NH_3, CO and N_2, the relative intensities of -CH_2- and -CH_3 groups of the 3.4μm absorption feature were found to change. It also resulted in enhancing the proportion of sp^3 bonded carbon. In the study where amorphous forsterite(Mg_2SiO_4) was prepared in the presence of methane gas, the absorption spectra of the material showed the strong 9.7μm silicate feature and also a feature at 3.4μm. This 3.4μm feature arises from the -CH_2- and -CH_3 groups that were incorporated in the silicate lattice. This 3.4μm feature is similar but not identical to the feature seen from carbonaceous materials.

2.6.4 Diffuse Interstellar Bands

The Diffuse Interstellar Bands are absorption lines detected from interstellar space. They are observed in the spectral range from 0.4-1.3μm with highest number of lines present in the visible region. They are too broad to be associated with atomic lines and are therefore called Diffuse Interstellar Bands (DIBs). Around 250 such lines are known at the present time.

The search of the carrier responsible for the observed diffuse interstellar bands, which has evaded an answer for a long time, has led to several laboratory studies of hydrocarbons. A discharge through Benzene and other hydrocarbons has resulted in the products of more than 65 new carbon chains and rings of astrophysical interest. A fairly new strong feature has been seen centered at 4429.27Å, which is close to strong interstellar band at 4428.9±1.4Å. But the band observed in the laboratory is significantly narrower than that of diffuse interstellar band. This could in principle arise due to the temperature difference between the two cases. The temperature in the supersonic beam experiment in the laboratory is around 2K, while the temperature in the diffuse interstellar clouds is higher $\sim$ 100-200K. The higher rotational temperature results in the excitation of larger number of rotational levels, which could make the band wider.

2.7 Optical Properties of Materials

2.7.1 Theoretical Considerations

The interaction of dust grains with electromagnetic radiation is a complex phenomenon which depends upon the nature, structure, shape, size and orientation of the particles. For quantifying the absorption and scattering of incident radiation, one generally defines a dimensionless efficiency factor for absorption(Q_{abs}) and scattering(Q_{scat}) and the efficiency factor for the total extinction which is the sum total of absorbed and scattered components i.e. $Q_{ext} = Q_{scat} + Q_{abs}$.

The albedo of the particle is defined as $\gamma = Q_{scat}/Q_{ext}$. The main aim is to calculate the efficiency factors for scattering and absorption. The theory of scattering by spherical particles of homogeneous composition was worked out by Mie (1908) and independently by Debye (1909). The evaluation of crosssection depends upon three factors, (a) the property of the material, generally specified by the complex refractive index, $\mu = n + ik$, where n and k refer to refractive and absorptive indices respectively. It can also be specified by the dielectric constant $\epsilon = \epsilon' + i\epsilon''$ of the medium, which is related to n and k through the relation $\epsilon' = n^2\text{-}k^2$ and ϵ''=2nk. (b) the wavelength of the incident radiation(λ) and (c) the size of the particle, a. However, in general, the quantity that enters in the calculation of crosssections is the parameter x, which is defined as

$$x = \frac{2\pi a}{\lambda}. \tag{2.6}$$

The theory of scattering for nonspherical particles, but for definite shape like cylinders, spheroids etc., have been worked out. In addition to the crosssections for scattering and absorption, the scattering intensity $S_1(\theta)$ and $S_2(\theta)$ as a function of the scattering angle θ can be calculated. Here $S_1(\theta)$ and $S_2(\theta)$ are the components of intensity in the direction perpendicular and parallel to the scattering plane. The scattering plane contains the incident radiation and the direction of the scattered wave. The phase function, $\phi(\theta)$, which gives the angular distribution of the scattered radiation, is known as scattering diagram, and can be calculated from a knowledge of scattering amplitudes $S_1(\theta)$ and $S_2(\theta)$. The other quantity of interest is the asymmetry factor for scattering, g, given by

$$g = \langle \cos\theta \rangle. \tag{2.7}$$

Here $\langle \cos\theta \rangle$ is the weighted mean of the cosine of the sacttering angle θ with the scattering intensity $\phi(\theta)$ as the weighting function. If the particles exhibit strong forward scattering, $\langle \cos\theta \rangle = 1$, whereas for isotropic scattering, $\langle \cos\theta \rangle = 0$.

However in a real situation, the particles are neither homogeneous in composition nor spherical in shape. For any shape of the particle, the method known as Discrete-Dipole Approximation (DDA) is generally used. In this method, the particle is composed of a large number of small sub units, each of which acts as a dipole oscillator. Through this method, it is possible in principle to take into account the fluffiness of the particle, surface roughness, voids, poracity etc. The inhomogeneous nature of the particles is generally represented by an average optical constant for the medium.

2.7.2 Laboratory Measurements

As can be seen from above, a knowledge of the optical constants of materials, namely the refractive indices or dielectric constants are essential for the calculation of efficiency factor for scattering and absorption. However they are not directly measurable quantities. They are generally derived from laboratory measurements of some property of light in combination with a suitable theory. The laboratory measurements could be the extinction from a slab, transmission or reflection of light as function of wavelength at near normal incidence or for various incident angles. These are then combined with theoretical formalism for extracting the optical constants of the medium. The most commonly used method for the determination of optical constants is the use of dispersion relations generally known as the Kramers-Kronig relations.

Basically, the dispersion relation is an integral over the whole range of photon energies which relate the dielectric function to the measured quantity like the extinction coefficients or the reflection measurements etc. The real and imaginary parts of refractive index μ or the dielectric constant ϵ are not independent of each other but are interrelated through these dispersion relations.

The determination of optical constants of materials from the laboratory measurements is quite difficult and cumbersome. The sample has to be perfectly smooth to avoid scattering effects. It should be homogeneous in composition which is difficult to achieve, as disorders are always present. In addition, the material is damaged due to exposure to high energy radiation. The imaginary part of the refractive index is also a function of temperature. Presence of minor impurities affect the results. It is also difficult to carry out the experiment over a wide range of wavelengths. The optical constants have been determined for several kinds of materials. For example, they have been measured for various types of amorphous and crystalline carbon, graphite, diamond, carbonaceous species, SiC, amorphous and crystalline silicates, oxides etc. They are also available for several terrestrial rock samples and moon samples. Few measurements refer to simulated space conditions. Quite often, the refractive indices of 'Astronomical Silicate' is used. This is a simulated material property and is derived directly from the astronomical observations of the 10μm silicate feature. The procedure adopted is to satisfy simultaneously the 10μm feature observations and the Kramers-Kronig dispersion relations, which involve optical constants of the medium. Therefore 'Astronomical Silicate' represent the average optical properties of the grain material.

The early reflection measurements on the graphite material leading to the determination of the dielectric constants of graphite, which when combined with Mie Theory of scattering for particles of sizes < 0.05 μm clearly had a broad feature at 2175Å, similar to the one seen in the interstellar extinction curve (Sec. 3.5.1.). A collective excitation of free electrons in graphite gives rise to this feature. This remains the most favoured explanation of the observed bump in the interstellar extinction curve even today.

Similarly, the first detection of 10μm silicate feature arising out of Si-O bond and later on the detection of 20 μm silicate feature arising out of O-Si-O bending mode from astronomical objects, prompted the measurement of the refractive indices of silicate materials. However, earlier observations of the 10 μm feature showed it to be smooth which led to the suggestion that it could arise from amorphous silicates. But observations carried out with ISO in the infrared region, revealed structures indicating the crystalline nature of silicates. In view of this, extensive measurements have recently been carried out on olivine and pyroxene for different Mg/Fe ratios. As an example, Fig. 2.16 shows the complex part of the refractive index k, for $Mg_xFe_{1-x}SiO_3$ for $x = 0.4$ to 1.0. The wide variation in the value of k

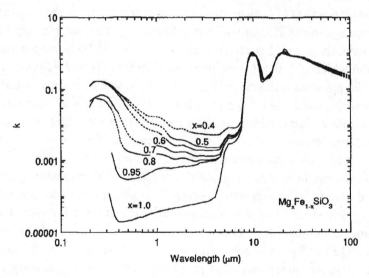

Fig. 2.16 Complex part k of the refractive index of pyroxene glasses, $Mg_xFe_{1-x}SiO_3$, for $x = 0.4$ to 1.0 (Dorschner, J., Begemann, B., Henning, Th., Jager, C. and Mutschke, H. 1995, Astron. Astrophys, **300**, 503).

with x can clearly be seen. This will have a great effect on the observable properties from such grains. The values of n and k for natural enstatite is shown in Fig. 2.17. The difference between the polarization directions are striking.

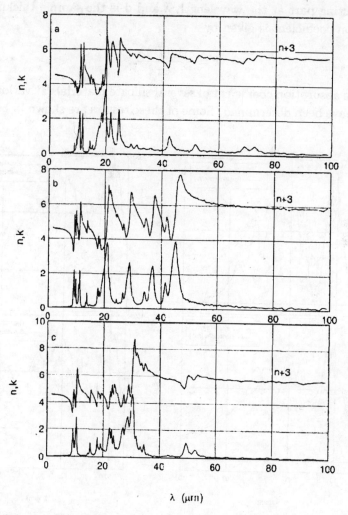

Fig. 2.17 Laboratory derived refractive index n and k for the three different crystallographic axes of the natural enstatite. (a) refers parallel to the c-axis and (b), (c) refer perpendicular to the c-axis(Jager, C., Molster, F.L., Dorschner, J., Henning, Th. et al. 1998, Astron. Astrophys., **339**, 904).

2.7.3 Mass Absorption Coefficient

For quantitative studies, the mass absorption coefficient k is required. Mass absorption coefficient gives a measure of the absorption of the material of the medium for the incident radiation. The mass absorption coefficient k can be determined from the analysis of transmission spectra. If T' is the corrected transmission for scattering due to roughness of the surface and reflecting part at the wavelength λ and d is the sample thickness, the absorption coefficient is given by

$$k = -\frac{\ln(T')\lambda}{4\pi d}. \qquad (2.8)$$

The mass absorption coefficient of several silicate materials at various wavelengths have been determined. Some of these results are shown in Fig. 2.18.

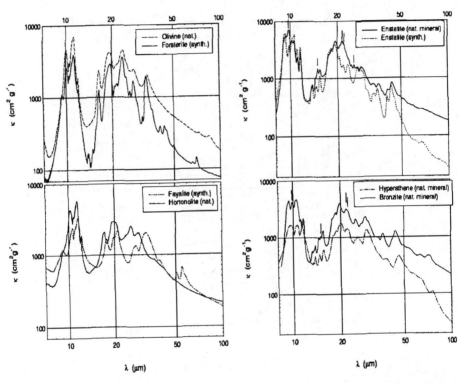

Fig. 2.18 Mass absorption coefficients calculated from transmission spectra of different olivine and pyroxene minerals (Jager, C., Molster, F.L., Dorschner, J., Henning, Th. et al. 1998, Astron. Astrophys., **339**, 904).

The detailed band positions and profiles depend on cation composition, i.e. Fe, Mg and other metals, chemically bonded into the silicate structure. Therefore, the changes in the spectral features with the variation in the Mg/Fe ratio has been studied. The peak position in the expected features as a function of Mg/Fe ratio for olivine is given in Table 2.5. The laboratory data shows a clear correlation between Mg/Fe ratio and band position. This result can be used to determine the iron content in crystalline dust.

Table 2.5 Peak positions in microns for different Fe/Mg ratios for olivines ($Mg_{2x}Fe_{2-2x}SiO_4$)

synthetical forsterite $x=1$	natural olivine $x=0.94$	natural hortonolite $x=0.55$	synthetical fayalite $x=0$	Comments
–	–	9.0sh	9.1m	
9.3m	–	–	–	
10.0s	10.0s	10.2s	10.4sh	asymmetric stretching ν_{as} of SiO_4
10.2sh	10.2sh	10.4?	10.6s	"
10.4m	10.5m	10.7m	10.9m	"
–	10.8w	–	–	"
11.2s	11.3s	11.3s	11.4s	"
11.9w	11.9m	12.0m	12.1m	symmetric stretching ν_s of SiO_4
–	–	13.2w	–	
13.6w	–	–	–	overtones
13.8m	–	–	–	overtones
14.6w	–	14.6?	–	overtones
16.3s	16.4s	17.0s	17.7m	asymmetric bending δ_{as} of SiO_4
18.3?	18.3w	19.1m	19.8m	"
19.5bs	19.5bs	20.5bm	21.0s	symmetric bending δ_s of SiO_4
20.8w	20.9w	–	–	rotation of SiO_4
21.5w	21.6w	–	26.6sh	"
23.5s	23.9s	25.5s	27.6bm	translation of one Me^{2+}
24.7w	25.3w	–	–	"
26.1m	26.4m	–	–	
27.5s	27.7s	28.4m	31.8bs	"
29.0?	–	–	–	
31.3m	–	–	–	"
33.5bs	33.8s	36.4w	39.5bw	"
36.3m	36.8sh	–	–	"
–	–	41.6w	–	"
40.7m	–	–	–	translation of one Me^{2+}
43.3m	–	–	–	
45.0w	–	–	–	translation of SiO_4
–	–	–	50.7w	translation of one Me^{2+}
49.8w	50.9w	51.8?	55.1w	translation of SiO_4
–	–	–	57.9sh	translation of one Me^{2+} & SiO_4
65.7w	–	–	–	translation of one Me^{2+} & SiO_4
69.7w	73.0w	84.4w	–	translation of one Me^{2+} & SiO_4
–	86.0bw	–	–	"

s, m, w, ?, b, sh refer to strong, medium, weak, doubtful, broad and shoulder respectively (Jager, C., Molster, F.L., Dorschner, J., Henning, Th. et al. 1998, Astron. Astrophys, **339**, 904).

The peak wavelength of silicate features shift to longer wavelengths with increasing Fe content. The usual indicator for the presence of Mg-rich olivine is the feature at 69.7μm, which is not present in iron-rich silicates. This feature shifts to 72.9μm for 10% Fe (i.e. $x = 0.9$). The crystalline olivine feature seen from astrophysical objects is at the expected non-shifted position of the band, indicating that they are Mg-rich. Similarly, pyroxenes are always found to be poor in iron. Therefore, crystalline silicates seen from astrophysical objects is Mg-rich and Fe-poor. The abundances of other elements is also low. This appears to indicate that the crystallisation process in astrophysical objects is highly selective. This could happen if by some process, metals other than Mg are selectively removed from the system. It could also happen if the process is less efficient for Fe-rich silicates. The band positions also depend upon cation composition, in particular, on Fe, Mg and other metals which are chemically bonded into the silicate structure. In general, the infrared spectra of pyroxenes are more complicated because polymerization leads to additional bands.

Therefore, in general the chemical homogeneity, particle size and shape distribution do have effect on the laboratory measured values. i.e. on peak positions and strengths. These will have direct effect on the identification as well as quantitative studies of the features seen in the infrared spectra of astronomical sources. Therefore, at the present time, many of the features have been identified with a fewer type of materials. However, lot of the features are unidentified due to lack of information on many other types.

2.7.4 *Microwave Scattering*

The scattering properties of non-spherical particles is a highly complex phenomenon dependent upon several factors. Hence, the computation of cross-sections have severe limitations. Laboratory studies relating to scattering and extinction by dust particles is a direct way of studying the complex behaviour of grains. They are also essential for checking theoretical solutions. Several measurements have been carried out on the extinction from a sample of sub-micron sized particles for testing the Mie Theory of scattering. However, one major problem associated with the particles generated in the condensation experiments is that they usually occur in aggregate forms, which may differ markedly from those of isolated particles. Therefore it is a major problem in comparing the laboratory measurements with those expected from single particles. The experiments carried out on carbon indicated a correlation between the width of the $\pi \rightarrow \pi^*$ absorption profile i.e. 2175Å feature, and the degree of clustering in the sample. The position of the feature was also found to be stable although the width of the feature increased with an increase in particle clustering. It is also not clear whether dielectric properties for the bulk material could be applied to particle of small sizes. The measurements have also to be carried out on carbon and silicates with a wide range in crystalline order.

The direct method of measuring the scattered intensity from single particles is rather difficult as the scattered intensities are generally too faint to measure accurately. So this method is suitable for a cloud of particles rather than single grains. This gives the average characteristic properties of the ensemble. The difficulties associated with this direct method can be overcome by using microwave analog method. The advantage of this method being, it is possible to scale the light-scattering problem up or down to any convenient dimension.

As mentioned earlier, the scattering properties depend on the parameter $x(=2\pi a/\lambda)$ and not individually on λ or size of the particle, a. In essence, light scattering measurements carried out on large size particles in the microwave region should provide the scattering properties of micron-sized particles at visual wavelengths. Microwave analog measurements have been carried out for compact, fluffy particles with or without absorbing mantle. These studies have provided a complete description of the scattering of these particles. The microwave analog technique is an important tool for testing mathematical models. The results of the studies agree reasonably well with those of calculated values, thus giving some credibility for the theories. However, one assumption involved in microwave analog studies is that the dielectric properties in the visual region is similar to the microwave region. This may not be the case in a real situation.

2.8 Microgravity Studies

The dust particles and aggregates in the universe are likely to be fractal aggregates (Sec. 2.10). They have fragile structure. These particles have been built in space where the densities are very low so that the gas flow is of the molecular type (Brownian motion). It is possible to produce low pressure in the laboratory experimental set up. But it is not feasible to simulate the Brownian motion even in a levitation compensated set up. Therefore, a reduced gravity environment appears to be the best place for conducting the experiment. A reduced gravity in the range of 10^{-2} to 10^{-6}g (g refers to the standard gravity of the earth) is quite similar to the dust in the universe with regard to velocity field and structure of the dust particles. There are several ways of producing the low gravity environment for short and longer time durations, starting from Aircrafts ($\sim 10^{-2}$g, for 10 to 30 secs) and going to Space Shuttle ($\sim 10^{-4}$-10^{-2}g, for 1 to 2 weeks) and Space Station ($\sim 10^{-4}$-10^{-2}g, for 1 month to a year). The choice of the method is dependent on the requirement of microgravity and the time duration for the experiment.

The scattering and polarization properties of levitating and aggregating dust particles have been studied using aircrafts. The microgravity environment is produced during quasi-parabolic trajectory of an aircraft, comprising between 30 to 35 parabolas. Several dedicated spacecrafts are operated on a regular basis by several space agencies and countries in the world for the study of aggregation of particles and their scattering properties The polarization studies have been carried out on various dust samples such as silicon carbide, boron carbide, basaltic glass and glass spheres with sizes ranging from 10 to about 200μm. There is reasonable agreement between the measured and the computed values except at large phase angles. This could possibly be due to depolarization induced by multiple scattering. The observed phase curves for irregular particles are similar to those seen from cometary or interplanetary dust particles. The values of polarization maximum are somewhat lower when measured under microgravity conditions than on the ground. This is consistent with higher values seen from packed grains compared to sifted grains. The results also indicate that compact dust particles show a blue polarization colour, while mixture of fluffy aggregates of submicron sized grains of silica and carbon, which agglomerate in highly porous structures, show a red polarization colour. The blue polarization suspected near the cometary nucleus, therefore imply the possible presence of large sized compact dust particles.

For the study of morphologies and growth rates of preplanetary dust, micron-sized SiO_2 particles were dispersed into a rarefied gas atmosphere(CODAG,Cosmic Dust Aggregation Experiment). Due to the absence of gravity, they are acted upon by the Brownian motion. The low velocity collisions($< 10^{-3}$m/s) between the particles lead to sticking and form aggregates with open structure. This experiment was designed for a shuttle flight.

The NASA'S Reduced Gravity Research Facility is a KC-135 aircraft (modified Boeing 707 jet) capable of flying in parabolic trajectories for producing short period reduced gravity environment. In an experiment carried out with this facility, the efficiency of vapour to crystalline solid growth of Zinc vapour, has been measured. Zinc was chosen as it is relatively easy to vapourize and also directly condenses into crystalline form. Zinc vapour was produced in a vacuum chamber containing argon gas. Vapour-phase transition is produced by cooling and the formation of crystalline zinc particles can be recorded. From the study of the increase in particle size distribution as a function of time, the sticking coefficient of zinc atoms was estimated. The results show that only a few of every 10^5 zinc atoms that collide with the grain surface sticks. This indicates that the direct deposition of vapour into crystalline grains is an inefficient process. It is not clear whether the results on zinc could be extended to astronomical materials and environments. If applicable, the large sized grains could not have been formed in the circumstellar outflows as the time scale for the formation of grains is much larger than the observed time scale of outflows. Therefore these grains could be formed in the stellar atmosphere prior to the outflow rather than in the outflow itself.

In an experiment carried out on the Space Shuttle, spherical Quartz particles (~1cm radius) were impacted into quartz sand and JSC-1 Lunar regalith simulant targets 2 cm in thickness. At impact speeds upto ~20 m/s, the injected particles got embedded in the target, while for higher speeds, the particles bounced back with significant amount of ejecta being produced. The velocity of the ejecta is found to be less than 10% of the impact velocity and the kinetic energy of the ejected material was less than 10% of the kinetic energy of the impacted particles. These results are of interest for planetesimal growth studies.

An experiment for the study of the behaviour of an ensemble of macroparticles charged by solar radiation was carried out under microgravity conditions onboard the Mir Space Station. The particles used for the study are Cerium Dioxide, CeO_2, Lanthanum Bromide, LaB_6 and spherical particles of Bronze covered by cesium. The selection of the particles was determined by their photoemission charging and low adhesion for lowering their sticking and precipitation on chamber walls. The particles are in a chamber with the buffer gas (neon) where solar radiation can impinge on them. The cloud of particles are made visible with a laser beam and recorded in a video camera. The results show that it is possible to form elongated, ordered structures of macroparticles.

Crystallization involve the growth process which also heats the crystal surface. The slower of the two processes, transport (mass) or kinetics (energy) in a situation, controls the growth of the crystal. Therefore crystallization could be different under microgravity conditions. For example, protein crystals grown in microgravity conditions were found to be most perfect obtained to date.

H_2O-ices is the most common type of material present in a wide range of objects in the universe. Therefore it has been studied extensively, both from experimental and theoretical means. The phase diagram of the H_2O molecule is highly complex, due to hydrogen bonding and proton disorder effects. Therefore, it can give rise to various types of crystalline solid phases, which is a sensitive function of temperature and pressure. Most phases are thermodynamically stable within a limited range of temperature and pressure. All the physical properties, such as density, conductivity, vapour pressure and sublimation rate of H_2O-ice, depend upon its crystalline structure. All the studies relating to ices in the universe is based on the laboratory studies of ice referring to 1g. However, it is not clear whether the laboratory data could be applied to ices present in the astrophysical environments. In addition, H_2O-ice could exist in a wide variety of forms in the universe, because of extreme range of physical conditions there, compared to the present knowledge on H_2O-ice, based on laboratory studies. Therefore the study of H_2O-ice in a low gravity environment (possibly with Space Station) is of interest for astrophysical studies as well as for atmospheric chemistry.

2.9 Nucleation

The formation of particles in the astrophysical environments is a complex phenomenon and is least understood. The processes involved are homogeneous nucleation, particle coagulation and surface growth. These are all functions of the physical conditions of the environment. Once the particles of critical size are formed, they collide with each other and coalese into bigger clusters leading to macroscopic particles. The nucleation process is the most difficult step in the whole scenario of particle formation.

Condensation of dust depend on temperature and pressure when some specific solid compound starts to condense from the supersaturated gas phase. This is achieved through the formation of cluster in the gas roughly with 10 to 100 atoms. The statistical fluctuations in the growth and decay of these precondensation clusters lead to some form of steady state size distribution. The classical nucleation theory based on thermodynamic considerations, was developed to provide theoretical framework for the calculation of rate of nucleation. The theory was developed basically to explain the earlier experimental results on vapour-liquid transitions. No data was available at that time, on vapour-solid transitions of refractory materials.

So it was not clear, whether classical nucleation theory could be applied to vapour-solid transition systems. In due course of time, several formalisms have been developed for the formation of grains for stellar environments, based on classical nucleation theory, which calculates the rate of nucleation in an expanding gas. However, the classical nucleation theory faces problems in applying to astrophysical environments. This comes about because of kinetic considerations, which will make it difficult for the formation of appropriate equilibrium precondensation cluster distribution, the very basis of classical nucleation theory. Further support for this conclusion comes from laboratory experiments.

Experiments have been carried out using several refractory materials for the study of conditions for the onset of homogeneous nucleation (Sec. 2.4.2.1). The main results that came out of these experiments is that, the critical size of the clusters formed were as small as dimers, metastable condensates were formed and the nucleation was found to be independent of the concentration of a thermodynamically condensable species (i.e. Fe or Mg silicates). Therefore, the results of experiments are in disagreement with the prerequisites for the classical nucleation theory. A better approach would be to use the detailed kinetic treatment which takes into account all the chemical reactions and processes involved. Several such methods have been developed.

2.10 Coagulation and Aggregation

When dust grains collide with relatively low velocity so that the individual submicron particles are not shattered or evaporated, the two particles can stick to each other leading to growth of the grain. This is generally termed as Coagulation and/or Aggregation.

Coagulation and Aggregation of submicron sized grains play an important role in a variety of astrophysical environments. This comes about from various considerations. Grain models based on porous or fluffy particles have been proposed to interpret interstellar dust observations. The dust particles present inside dense clouds appear to be larger in size and is attributed to the increased coagulation rate in such dense regions. The dust coagulation is quite rapid in the disks around young stellar objects, where the number densities are high. The micrometer size dust particles can agglomerate into the so called Planetesimals, which will eventually accrete into terrestrial planets. The direct evidence for the importance of such a process in the solar nebula comes from the study of Interplanetary Dust Particles, whose structure indicate clearly an aggregation of smaller particles with sizes around 0.1μm.

Several theoretical and numerical studies have been carried out for the study of physics of the process of coagulation between two colliding particles, in particular, the criteria for sticking as a function of particle size and collision velocities. These are based on the simple approach that when two particles approach each other the van der Waals force pull the grain together forming a contact area. This happens when the initial kinetic energy and the attractive interaction energy balances the repulsive elastic energy. Depending upon the magnitude of these two opposing forces, the particles can stick to each other or bounce back. Therefore it is possible to define a capture velocity v_{cr} such that if $v < v_{cr}$, the particles stick to each other and for $v > v_{cr}$, the particles bounce back. Another quantity of interest is the sticking efficiency ϵ, which is the probability of sticking for a given pair of dust grains colliding with velocity v. The two parameters v_{cr} and ϵ is a function of nature and structure of the dust particles. It is rather difficult to take into account all the effects in theoretical studies. Therefore laboratory experiments are important for a better understanding of the grain collision process.

Several dust coagulation and aggregation experiments have been carried out in the laboratory for the study of adhesion forces, sticking efficiency, capture velocity etc. In these experiments, particles of SiO_2 in the size range of 1μm were used. The adhesion force between micrometer sized particles has been measured directly. This is carried out by bringing into contact two grains of SiO_2 material of sizes 1μm and then measuring the force required to separate the two grains. The measured force was found to be consistent with the classical adhesion theory and are due to van der Waals forces. The collision behaviour of micrometer sized particles for different materials and morphologies in a broad velocity range upto $\sim$ 100m/s has been studied. The results show that spherical particles have rather sharp boundry for sticking at low velocities $\sim$ a few m/s and nosticking with increase in impact velocity. Therefore for collision velocities $\leq$1m/s the assumption of sticking efficiency of unity is reasonable. As the particle size increases the threshold velocity is found to decrease. The transition between sticking and nonsticking regions become smoother for irregular particles. The capture velocity for spherical particles is an order of magnitude larger than the theoretical estimate.

Experiments were also carried out for the study of morphologies and mass distribution function of the growing clusters for microsized grains of SiO_2 in a rarefied and turbulent environment. For this study, a cloud of deagglomerated dust grains was dispersed into a rarefied or turbulent gas environment. Frequent collisions among the deagglomerated dust grains leads to grain sticking and hence leads to rapid growth of dust aggregates. The aggregate morphology and mass distribution were then measured. The resulting aggregate was found to to be fragile and chain-like structure, which can be described by a fractal scaling law

$$m \propto S^{D_f} \qquad (2.9)$$

where m is the mass and S is the radius of the aggregate with a fractal dimension $D_f \approx 1.3$. The fractal dimension predicted by the models, give the values, $D_f = 1.95$ and 2.18 for aggregation with no restructuring and for maximum restructuring during impcts, respectively. As the velocity of impact is increased, the aggregates change their morphological structure and finally lead to the disintegration of the cluster.

The expected internal structure has been studied as the aggregation process takes place. This is generally represented by the filling factor, which is just the volume ratio of the dust grains to the total volume of the aggregate. This is derived from previously well characterized dust grains which have been deagglomerated into micrometer sized constituent grains in a rarefied gas with a typical velocity of around 0.1m/s. The particles are then allowed to deposit on a target leading to successive layers of dust grains. The measured filling factor is around $f \approx 0.15$. This implies that 85% of the volume is empty space and 15% of the volume is occupied by the particles. This indicate the high poracity of the dust grains.

Some studies have been carried out with iron particles of sizes around 0.02-0.05μm. With the application of magnetic field of around 100 Gauss, the iron particles were found to coagulate leading to structures like spiderweb, streamer and fluffy dust balls. Experiments have also been carried out on carbon grains condensed from a super heated carbon vapour. These grains are then electrostatically leviated during their growth. Several morphological types which resemble cauliflower-like structure present in meteorites (Fig. 5.6) as well as aggregates similar to interplanetary dust particles (Fig. 5.1), have been seen.

2.11 Other Dust Studies

Young Stellar Objects (YSO) are surrounded by circumstellar gas and dust and may be deeply embedded in molecular cores. X-rays can have effect on the grain composition. Very small grains may evaporate completely which suggest that the unidentified infrared features in the 3 to 13 μm region may disappear near X-ray luminous YSOs. This effect may have been seen in active galactic nuclei directly with ISO. Various materials relevant to interstellar grains (silicates and carbonaceous compounds etc.) have been exposed to X-rays from accelerator, to study the structural damage produced by X-radiation. They seem to indicate that dehydrogenation and breakage of aromatic rings in hydrocarbons may occur but have little effect on silicates. X-radiation will also cause photodissociation and chemical changes in the ices of dust grain mantles.

Experiments carried out with plasma-dust interaction has shown that plasma interaction leads to sputtering of molecules from the surface. These impacts also cause changes in structural property and alter surface chemistry.

In summary, the study of astrophysical dust analogs in the laboratory has helped enormously in our understanding of the nature, structure and composition of the dust particles that could exist under various astrophysical environments. However these studies are still limited as it is rather difficult to carry out the experiments for a wide range of astrophysical conditions. But serious efforts are being made to improve the situation. In essence, Laboratory Astrophysics, has a great role to play in the understanding of the cosmic dust.

References

A Good Reference Publication
Ehrenfreund, P., Krafft, K., Kochan, H. and Pirronello, V. (Eds.) 1999, *Laboratory Astrophysics and Space Research*, Kluwer-Academic Publishers.

A good account of infrared dust spectra is given in
d'Hendecourt, L.B. and Allamandola, L.J. 1986, Astron. Astrophys. Suppl., **64**, 453.
Pendleton, Y.J. and Allamandola, L.J. 2002, Astrophys. J. Suppl., **138**, 75.

Mixed ices is discussed in the following papers
Moore, M.H. 1999, In *Solid Interstellar Matter: The ISO Revolution*, Eds. L. d'Hendecourt, C. Joblin and A. Jones, Springer-Verlag, p.199.
Sandford, S.A., Allamandola, L.J. and Bernstein, M.P. 1997, From *Stardust to Planetesimals*, Eds. Y.J. Pendleton and A.G.G.M. Tielens, ASP Conference Series, Vol.122, p.201.
Schutte, W.A. 1999, In *Laboratory Astrophysics and Space Research*, Eds. P. Ehrenfreund, K. Kraffl, H. Kochan and V. Pirronello, Kluwer Academic Publishers, Dordrecht, p.69.

Silicate studies.
An excellent review on silicate studies is given in the following paper
Colangeli, L., Henning, Th., Brucato, J.R., Clemente, D., Fabian, D., Guillois, O., Huisken, F., Jager, C., Jessberger, E.K., Jones, A., Ledoux, G., Manico, G., Mennella, V., Molster, F.J., Mutschke, H., Pirronello, V., Reynaud, C., Roser, J., Vidali, G. and Waters, L.B.F.M. 2003, Astron. Astrophys. Rev., **11**, 97.

Other papers
Donn, B.D. and Nuth, J.A. 1985, Astrophys. J., **288**, 187.
Henning, Th. 1999, In *Asymptotic Giant Branch Stars*, IAU Symp. No.191, Eds. T.Le. Bertre, A. Lebre and C. Waelkens. p.221.
Thompson, S.P., Fonti, S., Verrienti, C., Blanco, A., Orofino, V. and Tang, C.C. 2002, Astron. Astrophys., **395**,705.

Photoluminescence, Silicon nano-crystals
Canham, L.T. 1990, Appl. Phys. Lett., **57**, 1046.
Delerue, C., Allen, G and Lannoo, M. 1993, Phy. Rev., **B48**, 11024.
Wilson, W.L., Szajowski, P.F and Brus, L.E. 1993, Science **262**, 1242.

Carbon studies can be found in the following papers
Frenklach, M. and Feigelson, E. 1997, In *From Stardust to Planetesimals*, Eds. Y.J. Pendleton and A.G.G.M. Tielens, ASP Conference Series, Vol.122, p.107.
Guillois, O., Ledoux, G., Nenner, R., Papoular, R and Reynaud, C. 1999, In *Solid Interstellar Matter: ISO Revolution*, Eds. L. d'Hendecourt, C. Joblin and A. Jones. Springer-Verlag, p.103.
Henning, TH. and Schnaiter, M. 1999, In *Laboratory Astrophysics and Space Research*, Eds. P. Ehrenfreund, K. Kraffl, H. Kochan and V. Pirronello, Kluwer Academic Publishers, Dordrecht, p.249.
Heymann, D. and Johnson, M.P. 2002, Astrophys. J., **574**, L91.
Sheu, S.-Y., Lee, I.-P., Lee, Y.T. and Chang, H.-C. 2002, Astrophys. J., **581**, L55.

Some studies relating to various aspects of PAHs are as follows
For a good summary, the following papers may be referred
Allamandola, L.J., Tielens, A.G.G.M. and Barker, J.R. 1989, Astrophys. J. Suppl., **71**, 733.
Puget, P.L. and Leger, P.L. 1989, Ann. Rev. Astron. Astrophys., **27**, 161.
Salama, F. 1999, In *Solid Interstellar Matter: ISO Revolution*, Eds. L. d'Hendecourt, et al. Springer-Verlag, p.65.

2175Å feature
Arnoult, K.M., Wdowiak, T.J. and Beegle, L.W. 2000, Astrophys. J., **535**, 815.
Duley, W.W. and Seahra, S. 1998, Astrophys. J., **507**, 874.

For Photostability refer to
Cottin, H., Moore, M.H. and Benilan, Y. 2003, Astrophys. J., **590**, 874.
Jochims, H.W., Baumgartel, H. and Leach, S. 1999, Astrophys. J., **512**, 500.

For Absorption Spectra
Moutou, C., Leger, A. and d'Hendecourt, L. 1996, Astron. Astrophs.,**310**, 297.
Sandford, S.A., Bernstein, M.P. and Allamandola, L.J. 2004, Astrophys. J., **607**, 346.

For Some papers on 3.4μm feature
Grishko, V.I. and Duley, W.W. 2002, Astrophys. J., **568**, L131.
Mennella, V., Baratta, G.A., Esposito, A., Ferini, G. and Pendleton, Y.J. 2003, Astrophys. J., **587**, 727.
Pendleton, Y.J. and Allamandola L.J. 2002, Astrophys. J. Suppl., **138**, 75.
Pendleton, Y.J., Sandford, S.A., Allamandola, L.J., Tielens, A.G.G.M. and Sellgren, K. 1994, Astrophys. J., **437**, 683.

For Diffuse Interstellar Bands, refer to
Allamandola, L.J. 1995, In *The Diffuse Interstellar Bands*, Eds. A.G.G.M. Tielens and T.P. Snow, Kluwer Academic Publishers, p.175.
Ball, C.D., McCarthy, M.C. and Thaddeus, P. 2000, Astrophys. J., **528**, L61.
Ruiterkamp, R., Halasinski, T., Salama, F., Foing, B.H., Allamandola, L.J., Schmidt, W. and Ehrenfreund, P. 2002, Astron. Astrophys., **390**, 1153.

The Theory of scattering can be found in these works
Bohren, C.F. and Huffman, D.R. 1983, *Absorption and Scattering of Light by Small Particles*, John Wiley, New York.
Debye, P. 1909, Ann. Physik, **30**, 59.
Hulst, van de. 1957, *Light Scattering by Small Particles*, John Wiley and Sons, New York.
Mie, G. 1908, Ann. Physik, **25**, 377.

Some Publications relating to optical properties are as follows
Draine, B.T. and Lee, H.M. 1984, Astrophys. J., **285**, 89.
Henning, Th, Il'In, Y.B., Krivova, N.A., Michel, B. and Voshchinnikov, N.V. 1999, Astron. Astrophys. Suppl., **136**, 405.
Jager, C., Dorschner. J., Mutschke, H., Posch, Th. and Henning, Th. 2003, Astron. Astrophys., **408**, 193.
Jager, C., Molster, F.J., Dorschner, J., Henning, Th., Mutschke, H. and Waters, L.B.F.M. 1998, Astron. Astrophys., **339**, 904.
Koike, C., Chihara, H., Tsuchiyama, A., Suto, H., Sogawa, H. and Okuda, H. 2003, Astron. Astrophys., **399**, 1101.
Mutschke, H., Anderson, A.C., Clement, D., Henning, Th. and Peiter. G. 1999, Astron. Astrophys., **345**, 187.
Taft, E.A. and Phillips, H.R. 1965, Phys. Rev., **138A**, 197.

For Microwave scattering study, see
Gustafson, A.S.Bo., Greenberg, J.M., Kolokolova, L., Stognienko, R. and Xu, Y.-L. 2001, In *Interplanetary Dust*, Eds. E. Grun, Bo. A.S. Gustafson, S. Dermott and H. Fechtig, Springer-Verlag, p.509.
Zerull, R.H., Gustafson, A.S.Bo., Schulz, K. and Thiele-Corbach, E. 1993, Appl. Opt., **32**, 4088.

The following papers refer to Microgravity studies
Colwell, J.E. 2003, Icarus, **164**, 188.
Ehrenfreund, P., Fraser, H.J., Blum, J., Cartwright, J.H.E., Garcia-Ruiz, J.M., Hadamcik, E., Levasseur-Regourd, A.C., Price, S., Prodi, F. and Sarkissian, A. 2003, Planetary Space. Sci., **51**, 473.
Fortov, V.E., Nefedov, A.P. and Petrov, O.F. 2001, Cosmic Research, **39**, 201.
Levasseur-Regourd, A.C., Cabane, M., Haudenbourg, V and Worms, J.C. 1999, In *Laboratory Astrophysics and Space Research*, Eds. P. Ehrenfreund, K. Krafft, H. Kochan and V. Pirronello, Kluwer-Academic Publishers, p.457.
Michael, B.P., Nuth, I.A. and Lilleleht, L.U. 2003, Astrophys. J., **590**, 579.
Worms, J.C., Renard, J-B., Hadamcik, E., Levasseur-regourd, A.C. and Gayet, J-F. 1999, Icarus, **142**, 281.

Some papers on nucleation
Egan, M.P. and Leung, C.M. 1995, Astrophys. J., **444**, 251.
Ferrarotti, A.S. and Gail, H.P. 2002, Astron. Astrophys., **382**, 256.
Wilcox, C.F. and Bauer, S.H. 1993, J. Chem. Phys., **94**, 8302.

Good review on Aggregation studies is given in the following paper
Blum, J. 2004, In *Astrophysics of Dust*, Eds. A.N. Witt, G.C. Clayton and B.T. Drain, ASP Conference Series, Vol.309. Astronomical Society of the Pacific Publications.

Chapter 3

Interstellar Dust

3.1 Introduction

The term interstellar matter or interstellar medium is generally used to denote the diffuse matter in the form of gas and dust which exists between stars. This matter forms a large fraction (∼20%) of the total visible mass of the Galaxy. Its presence is revealed only by the light of the stars which are modified due to the presence of this material. The samething is true for almost all the galaxies. Hence it will be useful to study in detail the interstellar dust in our Galaxy.

The present understanding of the life of the star is that, it is formed from a cloud of gas and dust due to gravitational collapse, star then goes through various evolutionary stages and finally the processed material inside the star is returned to the interstellar medium, which chemically enriches the original material. Hence stars forming out of this new material will be richer in heavier elements (metals), compared to the previous generation of stars. A fraction of these metals are incorparated into dust grains and the rest remain in the gas phase. The modification of the interstellar material has been going on through such a cycle for the last 12 to 15 billion years. In fact, the solar system material and the material out of which the life was created on Earth is basically this re-cycled material from the stars, which existed around 4 to 5 billion years ago.

Therefore, an understanding of the gas and dust components in the interstellar medium is an essential part of the understanding of the formation and evolution of stars and also the evolution of the Galaxy. The information about these components comes mainly from the interaction of electromagnetic waves with atoms, molecules and dust.

From the study of interstellar absorption lines, it is inferred that the matter existing in interstellar space is in the form of clouds. The spectra of interstellar lines taken at high spectral resolution show a large number of components, arising out of discrete clouds present along the line of sight. This arise due to clouds moving in different directions with different velocities giving rise to varying Doppler shifts which separate the components. This observation clearly shows that the interstellar matter is not homogeneous but exists in patchy cloud like forms. In fact, the morphology of the interstellar medium is highly complex, composed of several distinct regions with varying physical conditions. These are for example, Diffuse clouds, hot and warm clouds, molecular clouds and so on. The number density of hydrogen atom may vary anywhere from 1 to about $10^{4-9}/cm^3$ and the temperature may vary in the range from 10^4 to 10K or so. Diffuse clouds have low densities and Molecular clouds are characterised by very low temperature ($\sim$10K) and high densities ($\sim 10^{4-9}/cm^3$). Dust particles present in molecular clouds play a crucial role in the formation of highly complex molecules inside. The hot and warm clouds basically refer to their temperatures which are directly related to heating and cooling mechanisms by the gas and dust. These in turn will have direct effect on the pressure inside these clouds. The clouds contain in gas phase mostly hydrogen, about 10% helium and even less abundant other elements $\sim$1% of C, N and O. Other elements are even less abundant. Roughly 1% of the mass is contained in dust grains.

If stars with surface temperatures $\sim$ 30,000K are in the vicinity of these clouds, the radiation from the stars can ionize the hydrogen gas and create an emission region called Emission nebulae or Diffuse nebulae or HII region. These stars may be born inside the clouds and ionize the gas. However, if the temperature of the star is not high enough to ionise the hydrogen around it, the radiation of the star is just scattered by the dust particles in the cloud. These are known as Reflection nebulae. The nature and composition of dust particles in the interstellar medium can be derived from the studies of Diffuse clouds, HII regions, Reflection nebulae, Molecular clouds and so on.

Of course, direct evidence for the existence of interstellar dust comes from the photographs of our Galaxy. The conspicuous markings and patches in these photographs can clearly be seen (Figs. 3.1 and 3.2). These arise due to obscuring matter i.e interstellar grains. The next step is to understand more about the nature of interstellar grains. In particular, it is of interest to make an estimate of the amount of extinction (i.e sum of absorption and scattering), its effect on derived distances to astronomical

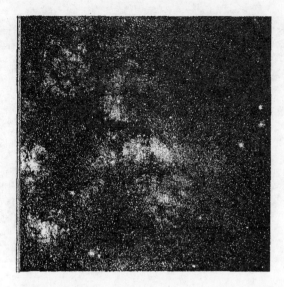

Fig. 3.1 Dark matter in sagittarius (Reprinted with permission of the Publishers From *The Milky Way*, by B.J. Bok and Priscilla F. Bok, 1957 Cambridge, Mass: Harvard University Press. Photograph taken by Ross with a 5-inch camera at the Lowell Observatory).

objects that lay in the background, the amount of dust and the composition of the dust.

In this chapter some of these aspects will be discussed. After a brief discussion of the effect of dust on the observations, the information derived from the interpretation of extinction, scattering and polarization measurements will be presented. This will be followed by infrared spectral studies. Here the observations of a few typical objects will be discussed to show the effect of environment on the dust particles.

3.2 Estimate of Amount of Extinction

When the radiation falls on a dust particle it can scatter as well as absorb the incident radiation. Therefore, the total amount of radiation screened by the dust particle in the line of sight is the sum total of the absorbed and scattered intensities. This is the total extinction by the dust particle.

The easiest method for estimating the interstellar extinction is to use the star counts. The method is to count the number of stars brighter than a certain given apparent magnitude in the obscured areas and compare them with the neighbouring relatively dust free regions. The assumption

Fig. 3.2 The Horse head nebula in Orion. The obscuring matter projects out in the form of the head of a horse (From *Exploration of the Universe*, G. Abell, 1964, Holt, Rinehart and Winston, Inc.,p.451).

involved in this process is that the distribution of stars in the two regions are practically uniform except for the dust causing the intervening absorption which is responsible for the observed differences. Figure 3.3 shows a plot of N_m, the number of stars per unit area brighter than certain visual magnitude,(m), versus m, for the two cases. The nature of the curves can be understood as follows: up to a certain interstellar distance corresponding to the point A, both the curves should behave the same way. For the case of absorption in the cloud, as one goes inside the cloud, the absorption is going to affect the star counts and so the curve deviates from the curve with no absorption. Beyond point B, the two curves should again behave the same way, except for the difference of a constant absorption, say $\triangle$m. Hence the two curves become parallel after the point B. However, in real situations, the sharp discontinuity in the two curves are not as drastic as shown in Fig. 3.3. This is because the absorbing material is more diffused than considered here. In spite of this, it is possible to get an estimate of the amount of extinction from such studies.

Another method is to use the Galactic clusters, which are bounded systems containing around 10^2 to 10^3 stars. One can make the assumption that there are a few typical standard types of Galactic clusters whose structures and sizes are independent of their locations and in that case these objects

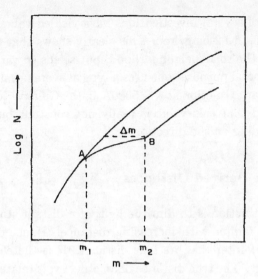

Fig. 3.3 The variation of star counts as a function of the visual magnitude for clear and obscured regions. Δm represents the absorption in the cloud.

can be used to get an estimate of the amount of interstellar extinction. By measuring the angular diameter α, of the cluster, the apparent linear diameter D, can be determined from the relation

$$D = \alpha r \qquad (3.1)$$

where r is the apparent distance. The apparent distance r, without absorption, can be determined from the study of several stars of the same cluster from the knowledge of their apparent magnitude m and the absolute magnitude M at $\lambda \sim 5000$Å, from the relation

$$5 \log r = m - M + 5. \qquad (3.2)$$

Such studies show that D was not constant as expected, but instead seemed to depend on the apparent distance r. This arises due to neglect of interstellar absorption. Therefore a modified relation of the following type is used

i.e. $\quad 5 \log r = m - M + 5 - Ar \qquad (3.3)$

where A is a constant referring to $\lambda \sim 5000$Å. It was found that $A \sim 0.8$ magnitude per kiloparsec (mag/kpc) was required to explain the observed discrepancy.

The results of several investigations have yielded values roughly in the range of about 0.4 to 2 mag/kpc. This clearly shows that the extinction is not constant in the solar neighbourhood, but seems to vary with location. However, for general purposes one uses a typical average value of 1 mag/kpc for the interstellar extinction at $\lambda \sim 5000$Å in the Galaxy. It should however be kept in mind that this value is really not constant but there is some variation depending on the direction.

3.3 Effect on Derived Distances

The effect of extinction is to dim the light of a distant star, and therefore make it appear fainter with increasing distance. Hence one might think that the star is fainter than its true distance. It is therefore necessary to correct this effect. The true distance of an object r, is related to its absolute and apparent magnitudes at any wavelength λ as

$$m(\lambda) = M(\lambda) - 5 + 5\log r + A(\lambda)r \qquad (3.4)$$

where $A(\lambda)r$ is the extinction term. If ρ denotes the distance of the object without the extinction term taken into account, then

$$m(\lambda) = M(\lambda) - 5 + 5\log \rho \qquad (3.5)$$

The relation between the true distance r and the apparent distance ρ is then given by

$$\log \rho = \log r + 0.2 A(\lambda) r. \qquad (3.6)$$

Figure 3.4 shows a plot of apparent versus true distances for several values of A for $\lambda \sim 5000$Å. It shows clearly that if the extinction is not taken into account, the estimated distances are generally overestimated.

It is interesting to note that with the availability of bigger and bigger telescopes, better instruments and techniques, it is possible to see fainter and fainter sources in the sky which are farther away, i.e one can probe deeper and deeper into space. This implies that the galactic system is much bigger in size than those of previous estimates. In short, the size of the galactic system increased every time better instruments or bigger telescopes become available. But on the other hand, the discovery of interstellar extinction had the opposite effect. The sources could be fainter not because they are farther away in distance, but due to the interstellar extinction

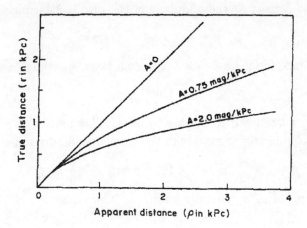

Fig. 3.4 The relation between the true distance(r) and the apparent distance (ρ) for two values of interstellar absorption (J. Dufay, 1968, *Galactic Nebulae and Interstellar Matter*, Dover Publications, 1968, p.151).

which absorbs the light of the source and hence makes it look fainter. In other words, the source could be bright and nearby, but looks fainter due to interstellar extinction. To a first approximation, one can use the average extinction coefficient to get an estimate for the distance to the object. This is one of the major uncertainties in the determination of distances to objects.

3.4   Amount of Absorbing Material

It is of interest to put an upper limit to the total mass of the interstellar medium composed of gas and dust. Such a limit was obtained by Oort in 1932 based on the following arguments.

Our Galaxy is a flattened disk and the stars and the interstellar material are moving in the gravitational potential of such a system. If the gravitational acceleration g(z) of the stars in the direction perpendicular to the plane of the disk can be measured, it directly gives the total mass of the material in the solar neighbourhood. The function g(z) can be deduced by measuring the population of stars in the z-direction of a homogeneous group of stars (like K-giants), whose velocities in the z-direction can be measured. This is related to the total mass density $\rho = (\rho_s + \rho_i)$, where ρ_s and ρ_i are the mass densities of the stellar and interstellar components respectively.

The total mass density at the midplane of the disc is determined to be,

$$\rho_s + \rho_i \sim 7 \times 10^{-24} g/cm^3 \tag{3.7}$$

From star counts, the mass density of known stars is estimated to be,

$$\rho_s \sim 3 \times 10^{-24} g/cm^3. \tag{3.8}$$

The difference is assumed to be due to interstellar matter. Therefore the total density of interstellar matter in the solar neighbourhood is

$$\rho_i \sim 4 \times 10^{-24} g/cm^3. \tag{3.9}$$

This upper limit of ρ_i is generally known as 'Oort limit'.

3.5 Nature of Dust

In order to get an idea of the nature of the material responsible for producing interstellar extinction, it is necessary to know the variation of extinction with the wavelength of the incident radiation. This curve is generally known as interstellar reddening or interstellar extinction curve. By comparing these results with those based on model calculations, it is possible to get an idea of the size and composition of the dust particles.

3.5.1 Mean Interstellar Reddening Curve

The simplest way to get the wavelength dependence of extinction (sum of absorption and scattering) is by comparing two stars of the same spectral type, one of which has plenty of material in front, while the other has no such material. The difference in their observed intensities with wavelength arises mainly due to the property of the intervening material.

If $m_1(\lambda)$ and $m_2(\lambda)$ represent the monochromatic magnitudes of the two stars of identical absolute magnitude $M(\lambda)$, one nearby and the other far off, then

$$m_1(\lambda) = M(\lambda) - 5 + 5 \log r_1 + A(\lambda) \tag{3.10}$$

and

$$m_2(\lambda) = M(\lambda) - 5 + 5 \log r_2 \tag{3.11}$$

where r_1 and r_2 are the distances of the two stars and $A(\lambda)$ is the extinction for the reddened star.

The difference in magnitudes between the reddened and the unreddened star is

$$\Delta m = m_1(\lambda) - m_2(\lambda)$$
$$= A(\lambda) - 5\log\frac{r_1}{r_2}. \quad (3.12)$$

The distance term $5 \log (r_1/r_2)$ can be eliminated by taking the difference in extinction at two wavelengths λ_1 and λ_2. The normalised extinction $E(\lambda)$ may be defined as

$$E(\lambda) = \frac{A(\lambda) - A(\lambda_1)}{A(\lambda_1) - A(\lambda_2)}$$
$$= \frac{\Delta m(\lambda) - \Delta m(\lambda_1)}{\Delta m(\lambda_1) - \Delta m(\lambda_2)}. \quad (3.13)$$

The value of $E(\lambda)$ can be calculated since $\Delta m(\lambda)$'s can be obtained from observations. This can be carried out at several values of wavelength. A plot of $E(\lambda)$ versus the wavenumber, λ^{-1}, is called the reddening curve or extinction curve.

Early observations were restricted to measurements in the visible spectral region of about 4000 to 7000Å. The portion A to B shown in Fig. 3.5 represents the resulting extinction curve for this spectral region. The shape of the extinction curve in the visible region is nearly the same for various regions of the Galaxy. This is called the mean extinction curve. It can be seen from Fig. 3.5 that the amount of reddening is inversely proportional to wavelength, for the visible region.

It is important to see how the reddening curve varies with the extension of wavelength base from the visible to ultraviolet and infrared regions. The extension to the ultraviolet region has been carried out with the use of rockets and satellites. Through such observations the reddening curve has been extended up to about $\lambda \sim 1000$Å. These studies gave very interesting results. The amount of extinction seemed to increase more or less linearly with λ^{-1} down to $\lambda \sim 2500$Å (Fig. 3.5). The observations also showed for the first time the presence of a conspicuous hump at $\lambda \sim 2200$Å ($\lambda^{-1} \approx 4.6\mu m^{-1}$), which can be seen from Fig. 3.6. Below 2200Å, the curve rises further. The observations for a large number of stars show that the hump occurs more or less at the same wavelength, $\lambda = 2175\pm 10$Å, independent of the region of observation. But it shows wide variation in width and strength. There is some indication that the feature appears to be broader in stars from dense clouds compared to narrower feature in stars from HII

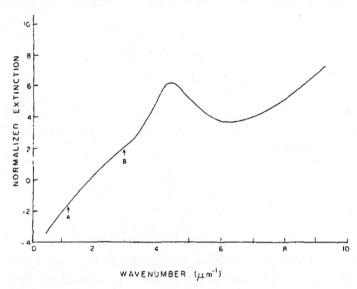

Fig. 3.5 Schematic representation of the mean interstellar extinction curve, which covers from far ultraviolet to infrared wavelength regions. The region A to B represents the curve based on ground based observations for spectral region of around 7000 to 3000Å.

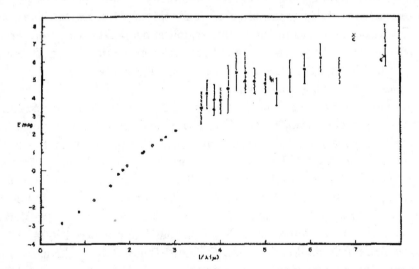

Fig. 3.6 Shows the Extinction curve, including rocket observations in the wavelength region 3000-1200Å. The feature around $4.5 \mu m^{-1}$ can be seen. (Stecher, T.P. 1965, Astrophys. J., **142**, 1683: reproduced by permission of the AAS).

regions. But the general nature of the curve is the same in the near and far UV, although there could be variation.

The reddening curve has also been extended to the infrared region ($\lambda \sim 10~\mu m$) using wide band infrared filters. The extinction curve seems to agree reasonably well with the extrapolation of the mean extinction curve. Some slight deviations have also been seen in certain cases. The shape of the reddening curve from the far ultraviolet to infrared region is shown schematically in Fig. 3.5, and represents an approximate mean reddening curve.

A quantity of interest is the total visual extinction A_v. Generally one defines the ratio R_v along the line of sight as

$$R_v = \frac{A_v}{E(B-V)} \qquad (3.14)$$

where E(B-V) is the colour excess in the UBV photometric system. R_v is basically an observational parameter measuring the slope of the extinction curve in the visible region (i.e $1/\lambda$ law). For a diffuse interstellar medium, R_v has nearly a constant value of around 3.1. For denser clouds the value of R_v show values $3.0 \leq R_v \leq 6.0$.

Having established the shape of the reddening curve, the next step is to identify the nature of the material and the sizes responsible for this type of behaviour. The candidate material so found has to satisfy the following requirements:

(1) extinction law varies as λ^{-1}
(2) average extinction coefficient ~ 1 mag/kpc at $\lambda \sim 5000$Å and
(3) mass density of absorber cannot exceed the Oort limit.

Several possible candidates given in Table 3.1 can be rejected straight-away as they do not satisfy the above requirements.

Table 3.1 Candidates for interstellar grains.

Particle	Law	Number density required to produce 1 mag/kps extinction
Free electrons (Thomson scattering)	independent of λ	300 cm^{-3}
Atoms or molecules (Rayleigh scattering)	λ^{-4}	10^4 cm^{-3}
Dust particles	λ^{-1}	

However, if one considers the scattering properties of electromagnetic radiation by dust particles of sizes(a) comparable to that of the wavelength of incident radiation(λ), it is found that the extinction produced varies as λ^{-1}. Therefore, the observed reddening curve can in principle be explained by particles of this type. The theory of scattering by particles of this type is a complicated problem and was developed by Mie in 1908 for homogeneous and spherical particles as mentioned earlier. The theory basically involves the solution of Maxwell's equations with appropriate boundary conditions. The dust particles present in the light path can absorb as well as scatter the incident radiation. These quantities are generally expressed in terms of dimensionless efficiency factors for absorption, scattering and extinction as Q_{abs}, Q_{scat} and Q_{ext} respectively. Therefore,

$$Q_{ext} = Q_{abs} + Q_{scat}. \tag{3.15}$$

To get cross sectional areas, these efficiency factors have to be multiplied by the factor πa^2, where a is the radius of the dust particle. The calculation of efficiency factors from the Mie theory depend upon the following three quantities: (1) the size of the particle, a (2) the wavelength of the incident radiation, λ and (3) the property of the material specified by the refractive index, μ=n+ik. Here n and k are the real and imaginary parts, representing respectively the contribution to scattering and absorption components. Figure 3.7 shows a typical curve for the variation of Q_{ext} as a function of the dimensionless parameter x=$2\pi a/\lambda$ for μ=1.33(i.e water). The curves are very similar in shape for any other refractive index. As shown by Fig. 3.7, Q_{ext} varies as λ^{-4} for small x and gradually changes over to λ^{-1} variation for range of x of about 2 to 5 with maximum around x $\sim$ 6. Therefore, the linearity of the reddening curve can be explained due to dust particles of sizes corresponding to this range of x.

3.5.2 Theoretical Extinction Curve

If τ is the optical thickness of the material between the star and the observer, then the observed intensity I is related to the original intensity I_0 by definition by the relation

$$I = I_0 e^{-\tau}. \tag{3.16}$$

Expressed in terms of magnitudes, it becomes

$$m - m_0 = 1.086\tau. \tag{3.17}$$

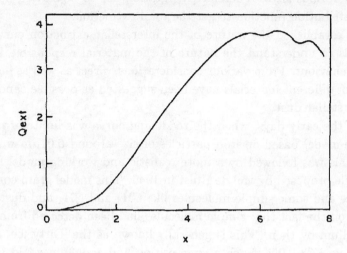

Fig. 3.7 The variation of extinction efficiency factor as a function of $x = 2\pi a/\lambda$ for m=1.33 and computed from Mie Theory.

The optical depth τ is given by the total extinction by the grains along the path from the source to the observer and can be written as

$$\tau = N\pi a^2 Q_{ext}(a, \lambda). \tag{3.18}$$

Therefore,

$$\Delta m = 1.086\tau = 1.086 N\pi a^2 Q_{ext}(a, \lambda) \tag{3.19}$$

where N is the column density, which is the number of grains in the column of 1 cm^2 cross section stretching from the star to the observer. It can be seen from the above relation that

$$\Delta m \propto Q_{ext}(a, \lambda). \tag{3.20}$$

Therefore, the normalised theoretical extinction curve can be derived from the relation

$$E(\lambda)_{cal} = \frac{Q_{ext}(\lambda) - Q_{ext}(\lambda_1)}{Q_{ext}(\lambda_1) - Q_{ext}(\lambda_2)}. \tag{3.21}$$

These curves can be compared with the observed reddening curve for assumed properties of the grain material.

It should be noted that in the above discussion, it was assumed that all the particles are of the same size. However in reality there exists dust particles of various sizes. Therefore integration has to be carried out over the size distribution function to get the total extinction.

Having established the nature of the interstellar extinction curve, the next step is to understand the nature of the material responsible for the observed behaviour. From various considerations, direct as well as indirect, a number of different materials have been suggested as possible candidates of the interstellar dust.

During the early days, when the reddening curve was limited to visual region, the model based on iron particles of size around $0.01\mu m$ was considered. This was followed by a more realistic and working model for the dust particle proposed by van de Hulst in 1946. This model grain consisted of water-ice and some simple molecules like CH_4 and NH_3 and dust. This was based on the fact that simple molecules had been detected from interstellar medium by then. This is generally known as the 'Dirty ice' model, which has $\mu=1.33-0.05i$. Such a composition and a$\approx 0.2\mu m$ could explain the observed reddening curve. It failed however, to produce the reddening curve in the UV region, when these observations became available. The detection of a hump at $\lambda=2200\textrm{Å}$ for the first time in the rocket observations in the UV region gave rise to the suggestion that the dust particles could be of graphite material, as this is the characteristic feature of graphite as seen in the laboratory. The large strength of the observed feature indicated that it must be produced by an abundant component of the interstellar gas. This requirement on band strength and relative abundance can be met with small carbon particles, such as graphite. However, the graphite grains fail to reproduce the far UV portion of the reddening curve. In addition, the observed central wavelength of the feature is found to be almost constant, but the width of the feature varies widely from one line of sight to another. These properties are not easily explainable from small graphite particles. Because of the cosmic abundance constraints, its great strength almost requires it to be produced by carbon in some form. The variability of the width of the feature can be explained in principle, by coatings of various kinds of materials on the small graphite grains, compositional inhomogenities, particle clustering etc. The constancy of the central wavelength of the feature require that all the graphite particles have about the same average size which is rather a severe condition. Other forms of carbon which have been proposed alternate to graphite are Platt particles which are unsaturated molecules of sizes $\sim 10\textrm{Å}$ or so, polycyclic aromatic hydrocarbons, Hydrogenated amorphous carbons, Quenched carbonaceous compound materials produced by laboratory discharges in methane plasmas, some form of coal, a kind of bacteria called E-coli and so on. These materials have a resonance near 2180Å and can also be produced more easily than graphite. Another component of the dust particle is the silicate material and there is strong evidence for this material to exist in interstellar space. This comes from the measurements made in the infrared region which showed the presence of a broad absorption feature at about 10μm. This spectral feature is the characteristic property of all silicate materials. Therefore various kinds of models are considered at the present time; bare grain or with coatings of different types of materials or a combination of two or three types of grains have been considered to explain the entire range of the reddening curve. Dust models based on aggregates, porous and fluffy particles with above materials have also been considered. However, 2175Å feature is more likely to be associated with some form of carbon type of material. The calculation of total exticton in the line of sight involves a knowledge of the size distribution of particles. The far UV rise in the extinction curve, indicated the presence of large number of small sized particles. The extinction curve in the visible require relatively larger size particles. The size distribution of the form n(a) $\propto$ a$^{-3.5}$ is found to reproduce the interstellar extinction curve in the wavelength range 1 to 9 μm^{-1} for spherical graphite and silicate grains with sizes between 0.005 to 0.25μm. This is frequently known as standard size distribution or MRN distributon for the dust in the diffuse interstellar medium.

The density of the dust particles required to explain the reddening curve in the visible region comes out to be

$$\rho_{dust} \sim 2 \times 10^{-26} \text{g/cm}^3. \tag{3.22}$$

Therefore the density of grains ~1% of the total density of interstellar material satisfies the Oort limit, even though its being far short of the latter.

3.5.3 Variations in the Interstellar Extinction Curve

So far, the discussion referred to mean interstellar extinction curve. This should represent the gross property of the grain material. However, deviations from the mean interstellar extinction curve have been seen from localized regions. The deviation from the mean extinction curve was first noticed in stars in the Orion nebula. This observation seem to indicate selective removal of small sized dust particles from the size distribution function. Such an effect can arise due to several physical processes, such as grain growth due to coagulation or destruction of grains, radiation pressure effect on small sized grains etc. Therefore, studies of interstellar extinction curve towards localized regions offer clear advantages over more general statistical analysis of large samples. The effects of specific characteristics of the region on the observed extinction can give clues to the deviation of the physical properties of the grains from the gross property of the grain. Moreover, since the initial properties of grains in the region were probably uniform any difference in extinction may be related to physical processes currently occurring in the region.

3.6 Interstellar Polarization

The light of the stars in the visible region was found to be polarized to the extent of a few percent ($\sim$ 3%). This was accidently discovered in 1948. The amount of polarization seemed to be proportional to the amount of interstellar extinction. The observed polarization arises mainly due to the preferential extinction by the dust grains, i.e., the extinction in the two transverse directions are different. This implies that the grains cannot be spherical in shape, but must be elongated. The grains must also be aligned to produce the observed polarization and this could be brought about by the weak interstellar magnetic field.

The wavelength dependence of polarization is shown in Fig. 3.8. It has a broad maximum in the visual range with its peak at λ_{max} varying from 3500 to 9000Å. The average value of $\lambda_{max} \sim$ 5500Å. The interstellar polarization for λ's<3000Å does not show any feature near 2175Å, but follows a steady decline. This is in contrast to high extinction present in this wavelength region.

The expected polarization from non spherical grains such as, infinite cylinders, spheroids etc. have been considered with MRN size distribution function for the particle sizes and with silicate material. Both models can provide a reasonable fit to observed interstellar polarization over the entire wavelength region. The derived axial ratio for oblate spheroids is around 1.4:1 for complete alignment case, to fit the observations. Grain models such as bare, core mantle and composite type have also been considered. The results based on bare model seems to be better.

In addition to the linear interstellar polarization, a weak circular polarization has also been observed. It is caused by the birefringence of the interstellar space due to the aligned elongated dust grains that produce

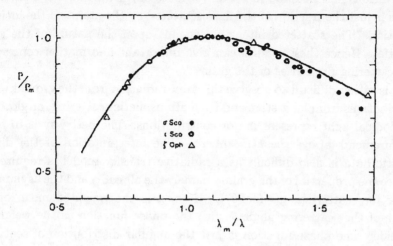

Fig. 3.8 Wavelength dependence of polarization for several stars compared with the empirical curve of Serkowski (Serkowski, K. 1973, IAU Symposium No.52 On *Interstellar Dust and Related Topics*, Eds. J.M. Greenberg and H.C. van de Hulst, D. Reidel Publishers, p.145: with kind permission of Springer Science and Business Media).

the linear polarization. Birefringence arises if the medium (crystal, grain etc) exhibit different properties in different directions for the incident light. This results in the plane of polarized light to rotate.

The linear and circular polarization observations taken together can put some restriction on the nature of the grain material. This comes about due to the fact that both types of polarizations are related through Kramers-Kronig dispersion relations which involve the refractive index of the material of the grain. The available observations indicate for the complex part of the refractive index to be ≤ 0.03. This shows the dielectric nature of the particles. This is also consistent with the detailed fit to the circular polarization data, which require $n=1.5$.

3.7 Scattered Light

3.7.1 *Diffuse Galactic Light*

The part of the incident radiation scattered by the dust particles should give rise to a diffuse radiation field in the Galaxy. This has also been observed in the visual region and is called the Diffuse galactic light. From these observations it is possible to get an estimate of the albedo of the particles, which is just the fraction of the light scattered by the grains to the incident radiation. The scattered intensity depends upon the nature of the grain material. Hence these studies can give important information concerning the scattering properties of the grains.

It is rather difficult to observe this faint radiation from the ground where contributions from faint stars and from atmospheric scattering, airglow and the Zodical light represent major contributions. The best way is to make measurements above the atmosphere. The interpretation of the diffuse galactic data is also difficult as a radiative transfer model is required to deduce the properties of the grains: namely the albedo γ and the asymmetry factor for scattering, g$(=\langle \cos \theta \rangle)$. Here, g is the weighted mean of the cosine of the scattering angle θ with the phase function as the weighting function. The phase function is just the angular distribution of scattered light. The radiative transfer involves a knowledge of the location of the scattering grains relative to the illuminating source. Such information is generally not available and hence involves making certain approximations. A phase function most commonly employed is the single parameter Henyey-Greenstein phase function given by the relation

$$\phi(\theta) = \frac{(1-g^2)}{(1+g^2-2g\cos\theta)^{3/2}}. \qquad (3.23)$$

The Diffuse galactic light has been obtained in the B band (λ_{eff}=4400Å) and R band (λ_{eff}=6420Å) from the measurements of the galactic background observations obtained by Pioneer 10 at heliocentric distances greater than 3AU. The deduced value for the albedo of the general interstellar dust is γ=0.61±0.07 and asymmetry factor g=0.60±0.22. Such studies have been extended into the ultraviolet regions based on satellite observations.

The Diffuse galactic light has been determined in the wavelength region 1500-4200Å from the instruments on board the Orbiting Astronomical Observatory OAO2. Both the latitude dependence of the diffuse galactic light at several wavelengths and the ratio of diffuse galactic light to direct star light have been measured in a large number of selected areas in the sky. Using these observations along with the radiative transfer model, the grain albedo and the asymetry factor has been derived. The results are shown in Fig. 3.9. The albedo show a sharp minimum at 2175Å coinciding with the peak in the interstellar extinction curve. This clearly points to the absorption nature of this band. The wavelength dependence of albedo and g factor derived from Diffuse Galactic Light observations is consistent with the values based on carbonaceous-silicate grain model. A major advantage of diffuse galactic studies is that the scattering characteristics derived for the dust represent an average value for the dust properties in our Galaxy.

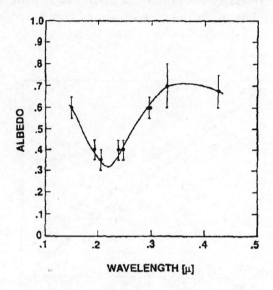

Fig. 3.9 The wavelength dependence of the albedo of interstellar dust derived from observations of the diffuse galactic light (Lillie, C.F. and Witt, A.N. 1976, Astrophys. J., **208**, 64: reproduced by permission of the AAS).

3.7.2 Reflection Nebulae

Reflection nebulae are generally associated with stars of spectral type B or later. These temperatures are not high enough to ionise the material (hydrogen) around it. Therefore they are made visible due to scattering by the dust particles. The extent to which reflection nebulae can be seen should depend on the temperature of the central star. Reflection nebulae are well suited for the study of interstellar grains as the matter viewed is typical interstellar material and also geometry of the nebulae is simple, as the exciting star is usually a single star of high luminosity. Information about the star's relative position with respect to the bulk of the nebulocity can in principle be obtained from the observation of stars reddening and the surface brightness gradient of the nebula in the visible light.

Hubble's early observations indicated that there existed a rather well-defined relationship between the apparent magnitude of an illuminating star (m_*) and the maximum angular distance from the star(a) at which nebula had sufficient brightness to be photographed (Fig. 3.10). The limiting brightness in the Hubble observation was 23.25 mag/square second of arc. The best straight line fit to these observations may be expressed by the relation

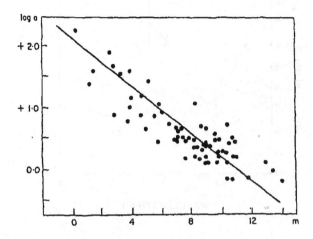

Fig. 3.10 Hubble's relation between the radius of a reflection nebula and the magnitude of the illuminating star (Hubble, E.E. 1922, Astrophys. J., **56**, 162 and 400: reproduced by permission of the AAS).

$$m_* - 4.9 \log a = 11. \qquad (3.24)$$

The interpretation of these observations require a model which involve among others, the properties of the dust particles. Based on a simple model in which it is assumed that the distribution of grains is uniform throughout the nebula and the phase function for the grains to be isotropic, the derived value for the albedo of the dust particles is $\gamma \geq 0.5$.

In order to fully exploit the potential of reflection nebulae, it is better to study individual bright objects in which accurate surface photometry over the entire nebular extent could be carried out. They are also easier to study as their geometry and the illuminating star is better constrained. Among many reflection nebulae that has been studied, Merope nebula and the nebula NGC7023 has attracted much attention. All these detailed studies have confirmed the high value of albedo in the visible ($\lambda \sim 5000$Å), $\gamma \sim 0.6$.

They have also shown the strong forward directed phase function g~0.6, expected from wavelength sized particles. Using the Ultraviolet Imaging Telescope of the Astro-1 mission, observations were carried out on the best studied reflection nebula NGC7023 in the wavelength region around 2800 to 1440Å. From a detailed comparison of the observed surface brightness distribution with those of model calculations the values for g and γ has been derived. The dust albedo in the near and far UV is, $\gamma=0.65$, which is as high as the albedo in the visible region. This indicates that optical properties are dominated by dielectric particles. The phase function is estimated to be g=0.75. The high value of g in the visible and UV regions indicate mainly forward scattering by grains, with sizes $\sim 0.1\text{-}0.5\mu m$. The albedo derived from several reflection nebulae in the region 2 to $8\mu m^{-1}$, agrees well with that of diffuse galactic light (Fig. 3.9) indicating the absorption nature of the 2175Å feature.

3.7.3 The Extended Red Emission

A broad, strong red emission feature around $\lambda \sim 7000$Å, was first detected from the reflection nebula in the Red Rectangle, in 1980. The feature was superimposed on the scattered continuum. The Red Rectangle is a highly symmetrical biconical nebulosity associated with the bright star HD 44179 (B9-A0III). The observed red emission feature is generally called the Extended Red Emission (ERE). Therefore the extended red emission is an intense, broad, structureless emission feature that appears between 5500 to 9000Å wavelength region and peaking around 6700Å with about 1200Å FWHM. The extended red emission is a common property of all reflection nebulae. It has also been seen in several other dusty environments such as, planetary nebulae, HII regions and external galaxies. The ERE can contribute as much as 35 % to the surface brightness of a reflection nebula in the R band($\lambda_{eff}=7000$Å). Laboratory studies show that many solids emit visible luminescence when exposed to ultraviolet or other energetic radiation. Therefore ERE is attributed to photoluminescence by some component of interstellar dust particles powered by ultraviolet photons. The fact that ERE is connected with interstellar grains comes about due to the fact that it is correlated with the dust column density and the intensity of the illuminating radiation. The observed spectral nature and photon conversion efficiency($\sim 10\%$) for UV and optical photons require the material to be abundant in the ISM, such as carbon and silicon-based material. The proposed carbonaceous material to explain the observed ERE are the same materials proposed for explaining the 2200Å bump in the interstellar extinction curve. In particular they could be Hydrogenated amorphous carbon, carbon nanoparticles, PAHs, fullerenes etc. The possibility that it could arise from silicon nano-particles in the size range 1 to 5nm has also been proposed.

The PAHs could be responsible for the observed ERE is related to the observation that AUIBs have been seen from objects where ERE is present. This indicates that the carrier responsible for both the emissions co-exist, but the same carrier is not necessarily responsible for the two emissions. This comes from the observation that the intensities of the two emissions appear not to be correlated. The difficulty associated with the PAH suggestion arises from the fact that photoluminescence arises mostly from neutral PAHs, incontrast to astrophysical environments where it is likely to be mostly in the ionized state.

The laboratory experiments have shown that Hydrogen Amorphous Carbon (HAC) emit photoluminescence in the visible region. The wavelength of the peak emission is a function of the degree of hydrogenation. The efficiency of photoluminescence is maximum when hydrogenation is maximum. But in this case, the peak of emission lies in the near UV region. The peak of emission could be shifted to visible region with decrease in hydrogenation. This however reduces the efficiency for photoluminescence by a large factor. Therefore to explain both the spectral range and the amplitude of ERE, the HAC model puts stringent condition on the degree of hydrogenation and also requires most of the carbon in the interstellar medium to contribute to photoluminescence process.

The expected photoluminescence from carbon nano- particles has been calculated. The model included carbon and hydrogen atoms with sp^2 bonds (i.e. aromatic bonds). The results of calculation predicted the presence of three features at 0.5, 0.7 and 1.0μm in the ERE spectrum. The 0.7μm feature can account for the observed feature in ERE. The feature at 0.5μm is produced by small sized particles. Such particles are not likely to be present in harsh environments. Hence, the feature at 0.5μm is not likely to be present in ERE. The other feature at 1.0μm has been looked for in two filaments in the reflection nebula NGC 7023 in the spectral region 0.35 to 2.5μm. Both the filaments exhibit strong features at 0.7μm but neither of them show a feature at 1.0μm. This casts some doubt on the carbon nanoparticle as the source of ERE.

The other suggestion that ERE could arise from photoluminescence of silicon nano-particles has also support from laboratory studies (Sec. 2.4.2.3.). Crystalline Si in the form of nano-crystals with different size distributions (sizes in the range 3 to 5μm) can explain the observed feature from various kinds of objects with regard to their spectral shape and position. The total mass of silicon nano-particles can be calculated from a knowledge of the measured photoluminescence efficiency of these particles, the reported efficiency of the diffuse interstellar medium dust component and the corresponding absorption crosssections. They indicate that the total mass of silicon nano-particles is roughly 1% of the total mass of the depleted interstellar Si atoms for explaining the observations. But this type of particle (oxygen rich) may have some difficulty in explaining the presence of ERE in C-rich planetary nebula.

3.8 Elemental Depletion

The elements in the gas and dust particles together represent the true composition of the interstellar cloud. If we assume that the elements in interstellar clouds have abundances represented by some standard composition then a measured deficit in the gas-phase abundance of an element must indicate that it is tied up in the dust particle. In order to carry out such an analysis it is necessary to have accurate gas-phase abundances for the interstellar cloud as well as a good knowledge of the standard composition. The standard composition is generally assumed to be solar abundances. The gas phase abundances of elements in the interstellar medium can place tight constraints on the possible types of interstellar grains that could exist.

The species that contribute most to interstellar grains are those with highest cosmic abundances such as, N, O, Ne, Mg, Si, S and Fe. Investigations of interstellar absorption lines show that dust forming elements are conspicuously underabundant in the interstellar gas.

The fractional depletion $\delta(x)$ of an element x is given by

$$\delta(x) = \frac{\left(\frac{N_x}{N_H}\right)_\odot - \left(\frac{N_x}{N_H}\right)}{\left(\frac{N_x}{N_H}\right)_\odot} \qquad (3.25)$$

where N_x represent the number of atoms of the element x and $\odot$ indicate the corresponding solar value. The fractional depletion can be defined as

$$\delta(x) = 1 - 10^{D(x)}. \qquad (3.26)$$

Here, D(x) represent the depletion index which is just the logarithm of the ratio of the interstellar to the solar abundance.

$$\text{i.e. } D(x) = \log\left(\frac{N_x}{N_H}\right) - \log\left(\frac{N_x}{N_H}\right)_\odot. \quad (3.27)$$

The value of $\delta(x)$ can vary between 0 and 1 corresponding to the cases, all the atoms are in the gas phase and all atoms are in the dust. From the study of interstellar absorption lines it is possible to get the column densities, N_x and N_H. Therefore the value of D(x) can be derived. Figure 3.11 shows a plot of the depletion index and the condensation temperature of the element. This temperature refers to the stage when significant amounts

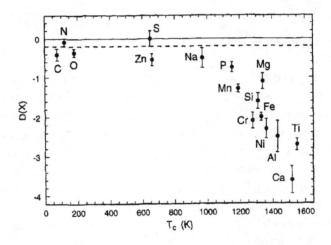

Fig. 3.11 Depletion index D(x) versus condensation temperature (Whittet, D.C.W. 2003, *Dust in the Galactic Environment*, Bristol: Institute of Physics publishers).

of elements should condense into solid form as the gas gradually cools below the respective T_c. As can be seen, the behaviour of the depletion exibits a good correlation with the condensation temperature of the elements. But the distribution pattern is quite distinct for T_c's, below and above $\sim$ 1000K. This is roughly the region of separation between the volatiles and refractory elements. Elements below 1000K show no depletions while for above 1000K, refractory elements show strong depletions. It shows the total depletion of refractory metals like Mg, Ca, Al, Si etc. This implies that the depletion of elements in the interstellar medium arises mainly because they are tied up in small solid grains.

The results as shown in Fig. 3.11 refer to physical conditions existing in diffuse clouds. Environment as well as the mean density of hydrogen, n(H), appears to have some effect on the depletion index D(x). The depletion of the elements is more for larger values of n(H). Therefore in denser regions where n(H) is large, accretion is favoured by the increased collision between gas phase species and grains. The depletion is more in warmer clouds compared to cooler clouds. Cloud velocities have also effect on the depletion of elements. Larger the velocity of clouds more is the depletion of elements.

3.9 Diffuse Interstellar Bands

Among interstellar absorption lines, a host of them are found to be weak and diffuse in character. They are called diffuse interstellar bands. From various considerations these bands are attributed to interstellar in origin. The lines do not share the periodic Doppler shift in binary stars but rather stationary in nature. The strength of these lines correlate with the amount of interstellar extinction, gas column densities and distance. The lines have been observed in the wavelength range of 0.4 to 1.3μm and exibit a diversity of profiles. Presently more than about 250 diffuse features have been identified. There could be many more but due to the weakness of the features and also being broad it may be difficult to detect them particularly when blended with stellar photospheric lines. The well studied feature is at $\lambda=4430$Å. Another feature of diffuse interstellar bands is the relative lack of features in the blue and near-ultraviolet. Even though the detection of such a feature was reported by Merrill as far back as 1936 till today the origin of the agent responsible for producing these features is not known. Several suggestions have been advanced to explain these diffuse features. But none of them is satisfactory. Broadly all the explanations fall into two categories. (1) They could be produced out of some simple or complex polyatomic spectra or a mixture of these. Studies carried out to identify the exact nature of this molecule or molecules have not been successful so far. (2) The other suggestion relates to the formation on grain surfaces. These studies have also not been very successful. The lack of success in finding the carrier of diffuse interstellar bands may be partly due to complexity of the interstellar processes, the wealth of diffuse interstellar bands and the potential carriers. In order to understand better, attempts are being made to put limits on the possible carriers from the observational point of view. These studies relate to the possible existence of correlation of strength of diffuse interstellar bands with variations in extinction curve parameters, grain properties, the interstellar radiation field, interstellar atoms and molecules and so on. Correlations among diffuse interstellar bands if present indicate that the same carrier could be responsible for all of them. The lack of strong features in the UV region can help in constraining the proposed carriers. This is an area which require some new insight into the problem. Here the laboratory experiments of the right type can help a great deal in elucidating this problem.

3.10 Infrared Spectral Features

3.10.1 *Diffuse Interstellar Medium*

The light from the star is both scattered and absorbed by the dust grains. The absorbed energy by the grains is transformed to heat. The grains assume an equilibrium temperature depending upon the properties of the dust grains in the UV, visible and near, far infrared wavelength regions. The temperature of the dust grains could be around a few tenths to a few hundred degrees kelvin depending on the environment. Earlier observations carried out from the ground and with rockets had detected this infrared radiation from various sources. These observations were later on followed by observations carried out with IRAS and subsequently by ISO satellite. The heating of the grains in the general interstellar medium is by the diluted diffuse stellar radiation field. The observations carried out with broadband passes at 60 and 100μm by the IRAS satellite detected distributed emission called IRAS Cirrus. This accounts for most of the galactic infrared background. The cirrus intensity at 100μm is found to correlate well with interstellar dust and gas. The observed far-infrared cirrus emission arise from the dust grains in thermal equilibrium with the interstellar radiation field. These results are consistent with the far-infrared measurements carried out with the COBE satellite. These observations showed the peak of emission to occur around 140μm. The estimated typical temperature of dust grains in the diffuse clouds in the solar neighbourhood is around 18K.

The B-type supergiant star No.12 in the Cygnus OB2 (VI Cygni) association (i.e Cygnus OB2 No.12) is the most luminous star in the Galaxy. Since it has a visual extinction of $A_v = 10$ magnitudes, it is well studied for the intervening dust of the interstellar medium. The major advantage of the study of this star being that it samples mostly the diffuse interstellar medium dust with negligible contribution of dust and ice from dense molecular cloud. The energy distribution of this highly reddened star is shown in Fig. 3.12. The conspicuous absorption feature of Si-O stretching mode of

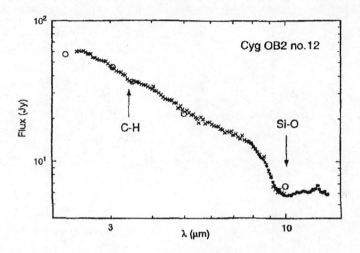

Fig. 3.12   Spectral energy distribution of Cyg OB2 No.12 (Whittet, D.C. and Tielens, A.G.G.M. 1997, In *From Stardust to Planetesimals*, Eds. Y.J. Pendleton and A.G.G.M. Tielens, ASP Conference Series, Vol.122, p.161: By the kind permission of the Astronomical Society of the Pacific Conference Series).

silicate can clearly be seen around 10μm. The lack of any structure in the silicate absorption feature is consistent with amorphous nature of silicates. Except for a weak absorption feature near 3.4μm no other feature seems to be present. The spectral observations of this star in the wavelength region 2.5 to 9μm has been obtained from Kuiper Airborne Observatory, Infrared Telescope Facility (NASA) and ISO. There is good agreement among all these measurements and show the 3.4μm to be weak. The 3.4μm feature has three subfeatures centered at 3.38, 3.42 and 3.48μm, as can be seen from Fig. 3.13. This 3.4μm feature has been studied in detail by comparing with the laboratory spectra of various materials to identify the exact nature of the dust. The laboratory spectra obtained from the residue of irradiated mixture of various ices such as (H_2O, CH_3OH, CO, and NH_3), (H_2O, CH_3OH, CO, NH_3 and C_3H_8) etc in different proportions, have been compared with the observed spectra. They indicate that the organic material in the diffuse interstellar medium is similar to the plasma processed pure hydrocarbon residue rather than to energetically processed ice residue. The three subfeatures at 3.38, 3.42 and 3.48μm are identified with symmetric and asymmetric C-H stretching frequencies in the -CH_2-(methylene) and

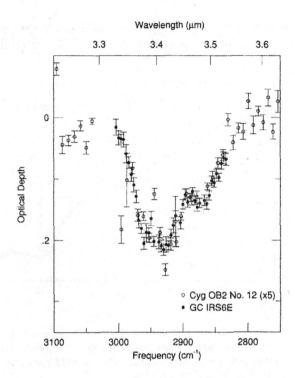

Fig. 3.13 Shows superposed CH stretch feature toward Cyg OB2 No.12 and the infrared source IRS6E near the Galactic centre (Pendleton, Y.J. and Allamondala, L.J. 2002, Astrophys. J. Suppl., **138**, 76: reproduced by permission of the AAS).

-CH_3(methyl) groups in aliphatic hydrocarbons, which have structures like -CH_2-CH_2-CH_3 and -CH_2-CH_2-CH_2-CH_3. The average ratio of -CH_2-/-$CH_3 \sim 2.5$. The 3.4μm spectra from hydrogenated amorphous carbon can also give a good fit to the observed spectra.

3.10.2 HII Region

The photographs taken in different regions of the Galaxy show some conspicuous patches of light or bright regions. They are nearly spherical in shape. These are known as Diffuse nebulae or HII regions or Emission nebulae or Gaseous nebulae. Among the bright emission nebulae, Orion nebula has been well studied. The bright regions seen in the photographs are actually the ionized interstellar gas. These are made visible through the emission lines from the ionized plasma. In this class of objects there are one or more stars which ionize the material surrounding it. The surface temperature of these stars are high $\sim$ 30,000K or so corresponding to spectral type B0 or O. Such objects are found to be confined to the galactic plane. The size of these nebulae are of the order of few parsecs.

Dust grains are also present in HII regions as can be seen from direct photographs. Many nebulae show dark patches and lanes indicating the absorption of light by the dust particles in them. The dust particles can also scatter the incident radiation. The scattered radiation has been seen from these nebulae confirming the presence of dust in these objects. The derived dust to gas ratio was normal in most of the cases except in the case of Orion nebula where a distinct variation appears to be present from the inner to the outer regions.

The observations in the infrared and far infrared regions have shown the presence of excess of radiation over and above the normal emission by the ionized gas of the HII regions. This is attributed to thermal emission from dust grains heated to temperatures $\sim$ 100 to 900K by the radiation of the star.

Earlier infrared observations in the Trapezium region of the bright Orion nebula had shown the presence of silicate emission feature at 9.7μm. This was later confirmed from the observations carried out with ISO.

Several HII regions have been observed with ISO in the wavelength region from 2.5 to 196μm. They show wide diversity in their spectral characteristics. The spectrum show smooth continuum with prominent dust features. In the mid infrared, the spectra show a series of well-known emission features at 6.2 and 7.7μm which are associated with C-C stretching and at 8.6 and 11.3μm which are associated with C-H bending modes of aromatics.

In the spectra of several HII regions embedded in dense clouds, the absorption features due to CO_2 and H_2O-ice have been observed. The CO_2 feature at 15μm shows detailed structure indicating the nature of the ice mixture. In particular the combination of polar and apolar ices can lead to sub structures. Detailed analysis of the observed spectra towards DR21 require a two component mixture, one component at 30K composed of H_2O, CO_2 and CO in the ratio of 100:8:8 and another component at 10K composed of H_2O and CO_2 with a ratio of 1:10.

3.10.3 Reflection Nebulae

As mentioned earlier these kinds of objects are made visible due to scattered radiation by the dust particles. Near infrared continuum measurements of several reflection nebulae in the region 1-5μm showed that the continuum was found to be far in excess over expected intensities of scattered light or thermal emission from dust particles in equilibrium with the radiation field. The continuum in the region 1-5μm indicated a high colour temperature of the order of 1000K. This component has been explained based on non-equilibrium thermal emission model from dust grains of very small sizes $\sim$ 10Å or so in which the grains are heated briefly to high temperatures by absorption of individual UV photons. The extension of observations to 13.5μm and later observations carried out with IRAS satellite showed that the very small grains contribution could be appreciable extending upto 20μm or so.

The bright reflection nebula NGC 7023 has been studied well in the optical and infrared regions. The nebula is illumunated by a star of temperature 17,000K (Be star HD200775). The short wavelength spectrometer spectrum of the reflection nebula NGC7023 from Infrared Space Observatory is shown in Fig. 3.14. Figure 3.14 show the presence of strong AUIB emission features. The feature at 16.4μm roughly corresponds to the feature in the resultant spectra obtained from the addition of 40 PAHs. Figure 3.14 show that the continuum under the AUIBs is very weak but increases for wavelengths larger than 20μm. The emission bands are rather thin and well separated. The infrared spectrum of NGC7023 as shown in Fig. 3.14 is very similar to the spectrum of several other reflection nebulae.

The spectra taken at the north west filament of the reflection nebula NGC 7023 in the wavelength region 0.35 to 2.5μm show two features at 1.15 and 1.5μm. The strong emission at 1.5μm has been attributed to β-FeSi$_2$ grains. The optical constants n and k of these grains have been determined in the far-infrared region. These are used to calculate the absorption behaviour of these particles in the far-infrared region. They indicate the presence of several bands in the region between 20 to 40μm. The detection of these features will give further support to the identification of FeSi$_2$.

3.10.4 *Molecular Clouds*

Molecular clouds are much denser than diffuse interstellar medium. Molecular clouds are characterised by high densities $\sim 10^{4-9}$ hydrogen atoms/cm^3. The sizes could be around 100pc. The dust contents in these clouds are quite high and therefore prevent stellar radiation field penetrating into the interior of the cloud. This leads to extremely low temperature in the cloud $\approx$10-30K. Molecular clouds exhibit internal structure and there is evidence

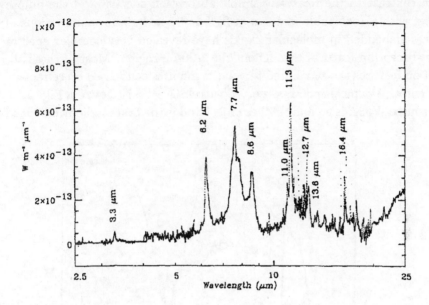

Fig. 3.14 SWS spectrum of the Reflection nebula NGC 7023 in the wavelength region 2.5 to 25μm (Moutou, C., Sellgren, K., Leger, A. et al. 1999, *Solid Interstellar Matter: ISO Revolution*, Eds. L. d'Hendecourt, C. Joblin and A. Jones, Springer-Verlag, p.89: with kind permission of Springer Science and Business Media).

for gravitational contraction. Therefore star formation can take place preferentially in such regions. The dust particles act as catalyst for the formation of complex organic molecules. Grain surface chemistry is quite active and leads to various kinds of molecules. Due to very low temperature and high densities existing in molecular clouds it is also the ideal place for the accretion of molecules onto dust particles leading to ice mantles. However as the star forms it destroys large quantity of volatile dust components due to heating, ionizing radiation and high velocity stellar wind. The dust mantles could also be transformed into organic refractories due to photolytic processing by the stellar UV radiation. Therefore the destruction, modification and condensation of the dust particles is a complicated function of the environment in which they are present. With the result the dust properties can vary from cloud to cloud or within the cloud depending upon their evolutionary status. Due to large dust extinction it is rather difficult to get direct information on the nature of the dust particles in dense molecular clouds. The information comes mainly from the observation carried out in the infrared region.

In contrast to the relatively simple absorption spectrum of the diffuse matter with recognizable features at 3, 3.4, 9.7 and 19μm, observations of sources embedded in molecular clouds have revealed very complex spectra with absorption features throughout the 3-20μm region. Most of these absorption features are attributed to volatile mantles condensed on refractory dust cores. A typical spectra observed with ISO-SWS is shown in Fig. 3.15. The source W33A is a young stellar object and is the best studied embedded

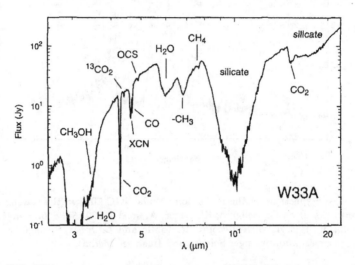

Fig. 3.15 ISO-SWS spectrum toward the protostar W33A. Several strong solid phase (ices) features can be seen (Gibb. E.L., Whittet, D.C.B., Schutte, W.A., Boogert, et al. 2000, Astrophys. J., **536**, 347: reproduced by permission of the AAS).

molecular cloud source. It shows almost all the strong absorption features to date towards molecular cloud sources. The spectrum is dominated by strong features of H_2O-ice and silicates centred around 3.05 and 9.7 μm respectively. The other prominant features are CO(4.67) and CO_2 (15.2). Other features present are H_2O (6.0), CH_3OH (3.53, 6.85), CH_4 (7.68) and HCOOH (5.83). The identification of solid methanol(CH_3OH) band at 3.54μm to the C-H stretches has been confirmed by extensive laboratory work on CH_3OH/H_2O mixtures. Further support for its identification came from the detection of other vibrational modes expected from solid methanol. The best evidence for the 4.9μm feature to C=O stretching vibrations of Carbonyl sulfide (OCS) came from the fit of OCS in a methanol matrix. Other mantle ice components are expected to be O_2 and N_2. However these apolar ices have very weak vibrational bands which are blended with other stronger features that makes it difficult to detect. The silicate feature arise from the core and the ices from the mantles.

The ISO-SWS spectra of a large number of various kinds of objects have been used to make a comprehensive study of the ice absorption features. They show rich spectrum with large number of strong and weak absorption features. Table 3.2 gives a listing of all these observed features with their wavelengths, identifications, approximate widths (FWHM) and band strengths. Table 3.2 show the complex nature of the icy material of the grain. The model fitting to the observed spectra has been carried out for all the sources studied based on laboratory studies of analogue material. As an illustration the results of study for the galactic center region (Sgr A*) is given in Table 3.6. Among all the species detected from various sources H_2O accounts for 60% -70% of the ice in most of the observed lines of sight. Figure 3.16 shows a comparison between the observed and the laboratory

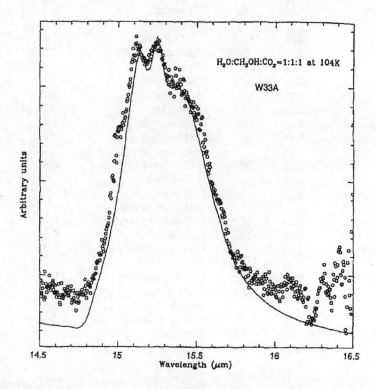

Fig. 3.16 A comparison between laboratory fit of the CO_2 bending mode feature at 15.2μm toward the proptostar W33A and an annealed ice mixture containing CO_2, CH_3OH and H_2O in equal proportions (Ehrenfreund, P. 1999, In *Solid Interstellar Matter: The ISO Revolution*, Eds. L. d'Hendecourt, C. Joblin and A. Jones, Springer-Verlag, p.231: with kind permission of Springer Science and Business Media).

spectra of the bending mode of CO_2 in W33A which is characterised by a triple peak structure at 15.15, 15.29 and 15.4μm. The laboratory data refer to ice mixture containing H_2O, CH_3OH and CO_2 with equal proportions.

The molecule CO is very efficient in freezing out in the coldest and

Table 3.2 Solid state features observed in the ices.

Molecule	λ (μm)	$\Delta\nu$ (cm^{-1})	Vibration Mode	A (10^{-17}cm molecule^{-1})	Ref.
NH_3 ···	2.96	45	-N-H stretch	1.1	1
H_2O ···	3.05	335	O-H stretch	20	1, 2
$-CH_2-, -CH_3$ ···	3.47	~10	C-H stretch	~0.1-0.4	1
CH_3OH ···	3.53	30	C-H stretch	0.76	1
CH_3OH ···	3.95	115.3	C-H stretch	0.51	1
H_2S ···	3.95	45	S-H stretch	2.9	3
CO_2 ···	4.27	18	C-O stretch	7.6	2
$^{13}CO_2$ ···	4.38	12.9	^{13}C-O stretch	7.8	2
H_2O ···	4.5	700	$3\nu_L$ and/or $\nu_2 + \nu_L$	1.0	2
"XCN" ···	4.62	29.1	CN stretch	~5	4
CO ···	4.67	9.71	^{12}CO stretch	1.1	2
^{13}CO ···	4.78		^{13}CO stretch	1.3	2
OCS ···	4.91	19.6	C-S stretch	17	5
H_2CO ···	5.81	21	C=O stretch	0.96	6
HCOOH ···	5.85	65	C=O stretch	6.7	7
CH_3HCO ···	5.83		C=O stretch	1.3	7
H_2O ···	6.02	160	H-O-H bend	0.84	1
$HCOO^-$ ···	6.33		C-O stretch	10	7
Organics ···	6.82	90	O-H bend, C-H deformation	1	1
HCOOH ···	7.25	16.8	C-H deformation	0.26	7
$HCOO^-$ ···	7.25	19.8	C-O stretch	0.80	7
$HCONH_2$ ···	7.22		C-H deformation	0.32	7
$HCOO^-$ ···	7.41	17.8	C-O stretch	1.7	7
CH_3HCO ···	7.41	10.6	C-H deformation	0.15	7
SO_2 ···	7.58	10-30	S-O (ν_3) asymmetric stretch	3.4	8
CH_4 ···	7.70	8	C-H (ν_4) deformation	0.73	8
CH_3OH ···	8.9	34	$C-H_3$ rock	0.13	1
NH_3 ···	9.35	68	Umbrella	1.3	9
Silicate ···	9.7	100	Si-O stretch		
CH_3OH ···	9.75	29	C-O stretch	1.8	1
H_2O ···	13.3	240	Libration	2.8	5
CO_2 ···	15.3	18	O-C-O bend	1.1	2
Silicate ···	18		O-Si-O bend		

Gibb, E.L., Whittet, D.C.B., Boogert, A.C.A. and Tielens, A.G.G.M. 2004, Astrophys. J. Suppl., **151**, 35.

densest molecular clouds. Polar ices dominated by H_2O-ice and aploar ices composed of molecules such as CO, O_2 and N_2 produce separate components in the absorption profiles of both CO and CO_2. In general the observed CO feature is a superposition of a narrow feature at 4.675μm and a broader feature at 4.681μm. Laboratory experiments have shown that the difference is due to matrix effect. The narrow CO feature is produced in the ice-matrix of predominantly apolar molecules such as N_2, O_2 and CO_2 while the broader feature arises in a matrix of polar molecules i.e H_2O. Polar ices generally evaporate around 90K under astrophysical conditions and hence can exist in high-temperature regions close to the star. On the other hand, apolar ices are highly volatile with evaporation temperature $\leq$ 20K and hence can exist in cold and dense region. Other species such as H_2CO, HCOOH, OCS and CH_4 are observed towards massive protostars and the observed abundances are of a few percent relative to water-ice. The abundances of various observed species is given in Table 3.3, which give a representative sample covering various types of objects. CO_2 is a common component of the ice mantle and is surprisingly present in large quantities $\sim$20% relative to water. However the abundance of gaseous CO_2 is quite low. The exact mechanism of its formation is not clear but it could be produced by grain surface reactions such as oxidation of CO or energetic processing by UV irradiation or cosmic rays or gas phase production in shocks and subsequent condensation on to the grains. The other molecule CH_3OH could be formed either by grain surface reactions or by energetic processing. The abundances of the species could vary from object to object depending on the several factors such as environment in which the stars are located, their evolutionary status etc. The abundance of CH_4 (7.68μm feature) is around 2 to 4% relative to water-ice.

The silicate spectral features at 9.7 and 19μm arising of core material of the dust grain are seen from young stars which are deeply embedded in the molecular clouds. These features are extremely wide without any profile structure or indications of crystallinity. Therefore silicate dust in molecular clouds appear to be amorphous in nature. However there is evidence that the crystalline dust grains are being injected into the interstellar medium by evolved stars. Therefore lack of crystallinity of dust grains in the molecular clouds indicate the amorphization of crystalline dust particles during their stay in the interstellar medium. This could have been brought about possibly by long time irradiation by cosmic rays.

Table 3.3 Column densities in solid phase toward RAFGL 7009S.

Molecule	Mode	Wavelength (μm)	N (cm^{-2})	% relative to H_2O
H_2O	ν_2	6.0	1.2×10^{19}	100
CO	ν_1	4.67	1.8×10^{18}	15
CO_2	ν_2	15.2	2.5×10^{18}	21
$^{13}CO_2$	ν_3	4.39	4.0×10^{16}	0.33
CH_4	ν_4	7.7	4.3×10^{17}	3.6
OCN^-	ν_2	4.62	4.4×10^{17}	3.7
H_2CO	ν_2	5.81	3.3×10^{17}	3.0
OCS	ν_1	4.9	2.0×10^{16}	0.17

Dartois, E., Demyk, K., Gerin, M. and d'Hendecourt, L. 1999, In *Solid Interstallar Matter: ISO Revolution*, Eds. L. d'Hendecourt, C. Joblin and A. Jones, Springer-Verlag, p.161: with kind permission of Springer Science and Business Media.

3.10.5 *Processes in Molecular Clouds*

The accretion time scale for dust particles in the molecular cloud is shorter than or comparable with both the time scales for reactions between ions and molecules which is $\sim 10^7$ years, and the average lifetime of molecular clouds which is $\sim 10^{5-7}$ years. Therefore at low temperatures existing in molecular clouds the impinging gas molecule should easily stick at grain surfaces. Some of these molecules will be sufficiently mobile at the grain surface, so that the chemical reactions can take place. H_2 formed on the grain surfaces can easily be desorbed from the grain surfaces. The surface reactions depend crucially on the H/H_2 ratio in the gas. If its value exceeds about 10^{-3}, O, C and N atoms will be hydrogenated to H_2O, CH_4 and NH_3, CO to H_2CO and CH_3OH and O_2 to H_2O_2 and H_2O. Otherwise CO, N_2 and O_2 should dominate the grain mantle ices. At high abundance of atomic oxygen, CO is oxidized to CO_2.

The collision of dust grains can lead to grain aggregates. The grain agglomeration occurs only for collisional velocity ≤ 10 m/sec. The properties of aggregates produced by agglomeration depend on the size and structure of the colliding particles. Very small grains (≤ 5nm) should form compact aggregates. For larger particles fluffy structure is expected. Thermal evaporation from grain surfaces is only possible at high temperatures. The sublimation temperature of ice mantles namely CO, CO_2 and H_2O, is vastly different and hence evaporate at different locations in the cloud which can produce local variation in the grain and gas composition. The dust mantle could also be processed by UV radiation, cosmic rays and so on. They can lead to sputtering, bond-breaking, implanting effects and so on.

3.10.6 Young Massive Protostar

The protostellar object RAF GL7009S is surrounded by heavily thick material containing cold grains. The infrared observations of this object is shown in Fig. 3.17. Strong solid state features of the ice mantles can clearly be seen. The extinction towards the source is so large that the water-ice stretching mode at 3μm, the CO_2 antisymmetric stretching mode around 4.27μm and the silicate stretching absorption band at 9.7μm are saturated. The spectra show ice absorptions arising from CO_2, CH_4, CO and others. The laboratory spectra shown in the Fig. 3.17 was obtained after UV photolysis of the initial ice mixture of the composition, $H_2O:CO:CH_4:NH_3:O_2=10:2:1:1:1$ deposited at 10K. The derived relative abundances of the molecules in the mantles is given in Table 3.3.

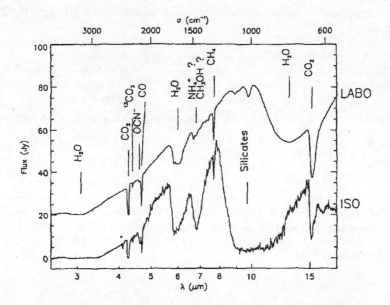

Fig. 3.17 A comparison between the SWS spectrum of RAFGL 7009S with the laboratory spectra (Dartois. E., Demyk, K., Gerin, M. and d'Hendecourt, L. 1999, In *Solid Interstellar Matter: The ISO Revolution*, Eds. L. d'Hendecourt, C. Joblin and A. Jones, Springer-Verlag, p.161: with kind permission of Springer Science and Business Media).

Gas phase abundances corresponding to ice constituents can also be estimated. This has been derived from the higher resolution observations of this object which reveal sharper transitions superimposed over the broad solid state absorption features. This makes it possible to derive the abundances in the gaseous phase and hence derive the gas to solid ratios for molecules. This is given in Table 3.4. Such studies are important from the point of view of understanding the mechanism of absorption, desorption, the role of grains in cloud chemistry etc. Table 3.4 shows clearly that CO_2 seems to be depleted very much in the gas phase compared to other abundant molecules present in the ice mantle. Laboratory studies have shown that CO_2 is produced very efficiently from the photolysis of H_2O and CO containing ices. This seems to indicate that the molecule CO_2 is produced mainly in the solid phase and reacts or destroyed when released in the gas. Another important aspect regarding H_2O is the location of their formation, whether on grains or in the gas. The formation of H_2O in the gas phase involves reactions with activation energies. Chemical models have indicated that the process of formation of H_2O in the gas phase and then accreting on the dust grains is highly unlikely. Therefore the formation of H_2O on grain surfaces is the mostly likely mechanism. In fact models show that water formation on grain surfaces is quite efficient and predict H_2O rich mantles.

Table 3.4 Observed gas to solid ratios in RAFGL 7009S.

Molecule	Gas (cm^{-2})	Solid (cm^{-2})	Gas/Solid
H_2O	$> 2 \times 10^{18}$ $< 1 \times 10^{19}$ (cold comp.)	1.1×10^{19}	≥ 0.18
CO_2	$1.0^{+1}_{-0.5} \times 10^{17}$	2.5×10^{18}	~ 0.04
CH_4	$1.2^{+0.5}_{-0.3} \times 10^{17}$	4.3×10^{17}	~ 0.28
CO	$6.1^{+1.7}_{-1.7} \times 10^{18}$ (cold component)	1.8×10^{18}	~ 3.4

Dartois, E., Demyk, K., Gerin, M. and d'Hendecourt, L. 1999, In *Solid Interstallar Matter: ISO Revolution*, Eds. L. d'Hendecourt, C. Joblin and A. Jones, Springer-Verlag, p.161: with kind permission of Springer Science and Business Media.

3.10.7 Star-Forming Regions

The first observation of the infrared star, known as BN/KL (Becklin-Neugebauer/Kleinmann-Low) source in the Orion nebula was reported nearly 30 years ago. This was the first observational evidence of a young star enshrouded in a dust cocoon. Since then many more young star forming objects in their parental clouds have been identified.

The formation of a star in the accompanying cloud introduces enormous change in the immediate environment. This has great effect on the gaseous and icy solid mantle composition of the dust grains. In particular, during collapse phase molecules accreting on to grain is quite efficient and the chemistry can be actively modified by surface reactions (especially hydrogenation of O, C and N to H_2O, CH_4 and NH_3 and oxidation of CO to CO_2) and possibly through photoprocessing of ices. After the formation of the star, its radiation will heat up the surroundings and the molecules from the mantles start evaporating, with the result the abundances in the gas phase increase. This process is likely to happen in sequence depending on the sublimation temperature of the species. All these lead to complex chemistry that could happen in the new environment. All these will also be a function of the evolutionary status of the star and the quiescent phase of the surrounding material. The derived gas to solid ratios for several sources are given in Table 3.5. The temperatures listed in the last column refer to

Table 3.5 Observed gas to solid ratio towards several objects.

Object	CO	CO_2	H_2O	CH_4	T_{warm} (K)
NGC 7538 IRS9	15	0.01	< 0.04	0.5	180
W33 A	100	0.01	0.02	0.5	120
GL 2136	200	0.02	0.4	> 1	580
GL 2591	> 400	0.04	~ 1	> 1	200-1000
GL 4176	> 400	0.04	~ 2	...	200-1000

van Dishoeck, E.F., Helmich, F.P., Schutte, W.A. et al. 1998, In *Star Formation with ISO*, ASP Conference Series, Vol. 132, Eds. J.L. Yun and R. Liseau, p.54, ASP Publishers).

those derived from the CO excitation. In all the sources CO is primarily in the gas phase, increasing toward warmer sources. CO_2 is mainly in the solid phase even in warmer sources. The gas to solid ratio for CH_4 is considerably higher than that of H_2O and CO_2 but lower than CO. These results show that as the young star evolves its radiation heats up the surrounding gas and dust. This results in more evaporation of the ice mantles resulting in enhanced gas to dust ratio. It may be noted that the sublimation temperatures of pure CO, CO_2, CH_4 and H_2O-ice under interstellar conditions are 20K, 45K, 20K and 90K respectively. The derived gas to solid ratios represent average values along the line of sight of the star. However in a real situation there could exist strong gradients in the abundance of these species both in the gas and on the grains in the surrounding medium. Figure 3.18 show in a schematic diagram the changes that can take place in the line of sight toward a massive protostar. In the region close to the

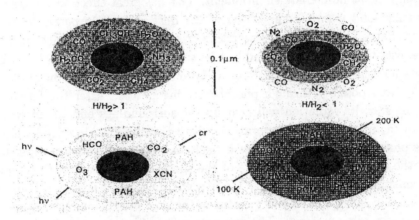

Fig. 3.18 Formation of polar and apolar in the protostellar environment. The simple ice species are converted to complex organic molecules, such as POM (Polyoxymethylene) and HMT (Hexamethyltetramine, $C_6H_{12}N_4$) under energetic processing and heating (Spaans, M. and Ehrenfreund, P. 1999, In *Laboratory Astrophysics and Space Research*, Eds. P. Ehrenfreund, K. Kraffl, H. Kochan and V. Pirronello, Kluwer Academic Publishers, p.1: with kind permission of Springer Science and Business Media).

protostar various energetic processes, such as strong UV radiation, shocks etc. could lead to high temperatures. The temperature should decrease moving away from the source. Therefore far from the central region, where the temperature is very low the ice mantle is mainly made up of say polar ice dominated by H_2O but also contain CH_3OH, CO, CO_2, CH_4 and other minor species. On the top of polar layer apolar ice dominated by CO, N_2 and some O_2 may accrete as an additional mantle layer. Moving from outward to inward region the temperature increases and this leads to evolution of ice mantles as trapped ices sublimate at specific temperature.

3.11 Galactic Centre

The central regions of the Galaxy is highly complex with various kinds of objects, clouds, non-thermal sources etc. There are also a wide variety of phenomena taking place on stellar to galactic scales. There is a compact radio source at its centre referred to as Sgr A* to differentiate it from the extended Sgr A complex in which it is located. The luminous objects situated at the Galactic centre provide background sources which can be used to investigate the interstellar extinction curve in detail at infrared wavelengths. The derived extinction is high, $A_v \sim 30$ mag. This makes it impossible to observe this region in the visual wavelength.

The abundance of heavy elements increases towards the centre of the Galaxy. This could result in more material to condense into dust grains and thus leading to higher dust-to-gas ratio in the inner Galaxy.

The atoms and molecules present in interstellar space give rise to absorption lines when viewed against a distant bright star. The strength of the absorption line depend upon the amount of intervening material. i.e extinction. Larger the extinction stronger will be the absorption line. Hence it will be possible to study the stronger absorption profile of the line in greater detail. This should provide the chemical composition of the grain material including the subgroups like $-CH_2-$, $-CH_3$ etc. Therefore observation towards Galactic centre with 30 magnitude of visual extinction should help in the study of interstellar dust component along the line of sight. The earlier observations in the wavelength region 8 to 13μm had clearly indicated the presence of a smooth 10 μm silicate feature. The ratio of the visible extinction to the silicate absorption depth near 10 μm along the line of sight to the Galactic centre is given by $A_v/\tau_{9.7} \sim 9$. This is about half of that found for the diffuse interstellar medium in the solar neighbourhood. This means that the silicate optical depth towards the Galactic centre is larger by a factor of two compared to the solar neighbourhood. This indicates the presence of more silicate dust grains towards the Galactic centre. This implies an increase in the M star (producing silicate grains) to carbon stars (producing carbonaceous grains) ratio in the inner parts of the Galaxy. This leads to an increase in the injection rate of silicate grains compared to carbon grains into the interstellar medium near the Galactic centre.

The 30 magnitude visual absorption towards the Galactic centre combined with the presence of bright infrared sources in the region have made the study of dust grains along the line of sight attractive. Studies of dust along the line of sight toward the Galactic centre Sgr A* have shown that it contains contributions from both diffuse and dense cloud dust. The ISO-SWS spectrum in the wavelength region 2.4 to 45 μm of the Galactic centre is shown in Fig. 3.19. The silicate features at 9.7 and 18μm are present. The spectrum towards Sgr A* has a rich infrared ice spectrum.

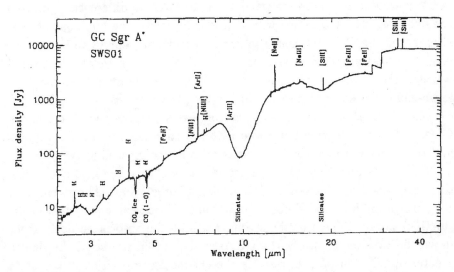

Fig. 3.19 SWS-ISO spectrum of the Galactic center (Lutz, D., Feuchtgruber, H., Genzel, R. et al. 1996, Astron. Astrophys., **315**, L269).

The absorption expected in ice grain material, notably the 3μm band of H_2O is present. Molecular cloud absorption features of solid state CO_2 (4.27,15.2μm) and CH_4 (7.68μm) are present. There is evidence for the presence of NH_3 (2.95μm) feature and possibly HCOOH. Laboratory studies for various ice mixtures have been used to fit the observed profiles of 3 and 6μm ice features. For this purpose the mid-infrared 2.4 to 13μm ISO-SWS spectra has been used. The best fit to the observed profile is obtained for the ice mixture of $H_2O:NH_3:CO_2$ (100:30:6, 15K) and HCOOH at 10K. The presence of aliphatic hydrocarbon dust is inferred from the detection of 3.4 (CH stretching mode), 6.85 and 7.25μm (CH deformation mode) features. The substructures present in the 3.4μm feature is identified with the CH stretch in methyl (-CH_3) and methylene (-CH_2-) groups. The average ratio of -CH_2-/-CH_3 $\sim$ 2-2.5. They are in saturated aliphatic hydrocarbons. The presence of aromatic hydrocarbons is identified via their 3.28μm (CH stretch) and 6.2μm (C-C stretch) modes.

Based on the ISO-SWS spectrum towards the Galactic centre region (Sgr A* source), all the strong and weak features have been identified. Table 3.6 gives a listing of all the observed features with their wavelength, identification and band strength. The column densities of the species are derived from the model fit to the observed spectrum based on laboratory studies of analogue material. As can be seen from Table 3.6, H_2O forms

Table 3.6 Spectral features present towards the Galactic centre source Sgr A*.

λ (μm)	FWHM (cm^{-1})	τ	Species	N ($10^{-17}cm^{-2}$)	$N/N(H_2O)$ *100	Ref.
3.0 ⋯	335	0.50(0.01)	H_2O	12.4(2.5)	100	1
3.28 ⋯		< 0.02	C-H stretch	⋯	⋯	1
3.46 ⋯	80	0.21(0.01)	HAC	⋯	⋯	1
3.54 ⋯	30	< 0.01	CH_3OH	< 0.5	< 4	1
3.95 ⋯	115.3	< 0.001	CH_3OH	< 0.4	< 3	1
4.27 ⋯	16.7	0.70(0.01)	CO_2	1.7(0.2)	13.7(1.6)	2
4.38 ⋯	7.2	< 0.07	$^{13}CO_2$	< 0.12	< 1	3
4.5 ⋯	700	[0.017]	H_2O	[12.4]	[100]	3
4.62 ⋯	29	∼ 0.14	XCN	∼ 0.83	∼ 6.7	4
4.67 ⋯	7.64		CO	< 1.5	< 12	1
4.91 ⋯	20	< 0.03	OCS	< 0.04	< 0.3	3
5.81 ⋯	21		H_2CO	< 0.3	< 2.4	1
6.02 ⋯	185	[0.07]	H_2O	[12.4]	[100]	3
6.0 ⋯		< 0.03	Organic residue	⋯	⋯	3
6.2 ⋯	60	0.05(0.01)	?	⋯	⋯	1
6.85 ⋯	88	0.05(0.01)	Organics NH_4^+	⋯	⋯	1
7.243 ⋯	19	0.03(0.01)	HCOOH	0.8(0.2)	6.5(1.6)	1
7.676 ⋯	10.4	0.017(0.003)	CH_4	0.30(0.07)	2.4(0.56)	1
9.0 ⋯	68	< 0.012	NH_3	< 0.61	< 4.9	5
9.7 ⋯	176	2.31(0.02)	Silicate	⋯	⋯	3
13.3^a ⋯	240	[0.11]	H_2O	[12.4]	[100]	3
15.2 ⋯	20.3	0.077(0.005)	CO_2	1.7(0.2)	13.7(1.6)	2
			(Polar)	1.7	13.7	2
			(Nonpolar)	< 0.1	< 0.8	2
18 ⋯	140	0.57(0.01)	Silicate	⋯	⋯	3

Gibb, E.L., Whittet, D.C.B., Boogert,A.C.A. and Tielens, A.G.G.M. 2004, Astrophys. J. Suppl., 151, 35.

the major component of the icy material towards Galactic centre and also show the complex nature of the icy material.

3.11.1 Extinction Law

The extinction law towards the Galactic centre has been deduced from the study of hydrogen recombination lines in the wavelength region between 2.5 and 9μm. This is derived from a comparison of the observed fluxes with the expected fluxes based on model calculations. The resulting extinction curve is shown in Fig. 3.20. It shows a flat distribution in the wavelength region 4-8μm in contrast to a pronounced minimum expected for standard graphite-silicate mixture corresponding to the general interstellar extinction

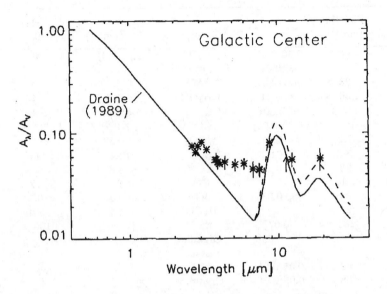

Fig. 3.20 The derived extinction law towards Sgr A* from SWS-ISO hydrogen recombination line data (Lutz, D. 1999, In *The Universe as Seen by ISO*, Eds. P. Cox and M.F. Kessler, ESA-SP 427, p.623).

curve. There could be additional absorptions which makes the curve flatter. However the silicate features are clearly present.

Several studies have been carried out with regard to the evolution of gas and dust phases of the interstellar medium in our Galaxy based on standard models for its dynamical and chemical evolution. In particular it takes into account the dynamical infall model for the evolution of the stars, chemical evolution models for the evolution of elements and a model for the evolution of dust in which dust is formed in quiescent stellar ejecta and supernova of Type I and II ejecta and is destroyed by expanding supernova blast waves. Accretion inside molecular clouds allows for grain growth. Such models can reproduce reasonably well several observational results such as, the relative and absolute elemental abundances at the time of formation of the sun, the age-metallicity relation, Galactic iron to hydrogen abundance gradient and so on. The silicate and carbon dust rates depends strongly on stellar mass with carbon dust produced primarily in low-mass carbon stars. The delayed recycling of the gas and dust by these stars back to ISM has important consequences for the evolution of the dust composition in various regions of the Galaxy. The evolution of ISM metallicity and those of the carbon and silicate dust are quite similar. The relative abundance of carbon to silicate dust depends on the birthrate history of the gas and evolves as a function of time. There could be three phases in the evolution of the dust composition. In the first phase the dust population is silicate rich since carbon stars have not yet evolved off the main sequence. In the second phase these stars inject carbon-rich dust into the ISM and the carbon to silicate mass ratio increases significantly. In the third phase, the lowest mass carbon star evolves off the mainsequence and is marked by an increase in the abundance of silicate dust. These phases depend upon region of the Galaxy since it depends upon the birthrate history of the gas in that region. The evolution of dust abundance and composition plays an important role in determining the extinction law in the Galaxy and in external galaxies. Young galaxies or star forming regions that underwent a recent burst of star formation will be silicate rich and their extinction law may be characterised by the absence of 2175Å extinction bump generally attributed to carbon grains. The Small Magellenic Cloud may be such a galactic system. Our Galaxy extinction law belongs probably to the second phase. In the third category there is excess of carbon dust as compared to the Galaxy and may be present in galaxies or star forming regions in which most massive carbon stars are evolving off the main sequence.

In brief the dust particles ejected from AGB stars are subjected to extended periods of ultraviolet and ion radiation in the interstellar medium. This could modify the chemical and physical structure of original grains. On the other hand, the dust grains present in molecular clouds can grow thick mantles of organics around the original grains. When a massive star is formed in a molecular cloud it changes the environment drastically which has a great effect on the dust grains. Therefore the nature, structure and composition of dust particles in the interstellar medium is complex involving star, gas and dust and the modifications produced due to the hostile environment in which it is embedded.

3.12 Sources of Dust

The basic problem regarding interstellar dust is their origin. The idea that the nucleation of grains from low density interstellar gas clouds was proposed by Lindbland in 1935. This idea subsequently led Van de Hulst to suggest in 1949 'Dirty Ice' model with molecular composition made up of abundant elements like C, O, N and H, embedded in an icy-matrix. It was thought that the dust particles could be produced in interstellar space through molecular collisions. i.e the simple molecules combine to form complex molecules progressively which finally lead to formation of dust particles. However it is highly unlikely that the dust particles could be formed by this process in interstellar space where the density is low and so the collisions are very infrequent. Hence the formation of dust particles in interstellar space is highly unlikely. One has to look for regions much denser than in interstellar space where particles can nucleate and grow. It could in principle be formed in the atmospheres of cool stars, nova, supernova ejecta and so on. The difficulty of molecule formation and nucleation encountered in interstellar space is overcome as the photospheric density in a giant star $\sim 10^{15}$ to 10^{16} atoms/cm^3. At such high densities collision between atoms are very efficient which can lead to nucleation and growth of grains. These dust particles are then dispersed into the interstellar medium through mass loss process. The dust injection rate into the interstellar medium depend upon several factors such as Galactic distribution of mass-losing stars, the average mass loss rates for these stars, the types of dust particles that can condense in the outflows and the dust-to-gas ratio, in the outflows. The estimated injection rate of dust into ISM is given in Table 3.7.

As can be seen from the table the bulk of the dust particles originate in oxygen-rich M giants, OH/IR stars and supergiants producing dust rich in silicon, metals and oxygen and in carbon-rich stars producing carbonaceous type of particles. There could be large uncertainty in the contribution of dust from Supernovae of Type II to the interstellar medium. Planetary nebulae, novae and Wolf Rayet stars provide only a small fraction of dust. Table 3.7 show the minimum dust formation rate $\sim 8 \times 10^{-6}$ M$_\odot$kpc^{-2} yr^{-1} averaged over the Galaxy and maximum dust formation rate $\sim 3 \times 10^{-5}$ M$_\odot$ kpc^{-2}yr^{-1}. With an injection rate for gas of 10^{-3} M$_\odot$kpc^{-2}yr^{-1}, and assuming that the interstellar dust mass is 1% of the gas mass (i.e 0.01 $\times 5 \times 10^9$M$_\odot$), the interstellar dust can be replenished in a time scale $\sim 3 - 6 \times 10^9$ yr.

The dust in the ISM could consist of a mixture of dust particles condensed in circumstellar environments and those modified in dense clouds. The dust particles of the first category retains their memory of its site of formation such as elemental or isotopic ratios characteristic of the nucleosynthetic processes occuring within particular stellar interiors. The examples of this type are the interstellar dust particles of SiC and graphite recovered from meterorites. On the other hand the modified dust retains no memory of its formation in a given stellar environment as they will have

Table 3.7 Contribution of various sources to interstellar grains.

Source	Contribution (10^{-6} $M_\odot$ kpc^{-2} yr^{-1})		
	carbon	silicate	silicon carbide
C-giants	2	–	0.07
M-giants	–	3	–
novae	0.3	0.03	0.007
planetary nebulae	0.04	–	–
supergiants	–	0.2	–
WC stars	0.06	–	–
supernovae II	2	12	–
supernovae Ia	0.3	2	–

Jones, A.P. and Tielens, A.G.G.M. 1994, *The Cold Universe*, Eds. T. Montmerle, C.J. Lada, I.F. Mirabel and T. Tran thanh van, Gif-sur Yvette: Editions Frontieres, p.35.

been processed in interstellar clouds leading to mantle growth and so on. It is likely that the dust particles of this type are the GEMS (Glass with Embedded Metal and Sulphides components) seen in meteorites.

3.13 Detection of Interstellar Dust:*in-situ*

3.13.1 *Spacecraft Studies*

The evidence that interstellar grains are continuously entering the solar system comes from *in-situ* measurements of dust by spacecrafts. The objective of dust experiments from spacecrafts is to study their physical and chemical behaviour and also determine their orbit which would help in understanding the source of these dust particles. The *in-situ* measurements of dust particles have been carried out for heliocentric distances of 0.3 to 18AU. The limitation of the dust detectors in the spacecraft is that the particles have to reach the position of the detector and also hit the sensitive area for its detection. Therefore the range of orbital elements of particles to be detected are limited. Of particular interest here is the detection of interstellar particles passing through the planetary system by the Ulysses spacecraft. The identification of the observed dust particles as due to interstellar dust comes out of several considerations. At Jupiter distance the orbit of these particles are predominantly retrograde in contrast to those of prograde direction of interplanetary dust particles and also coincides with the flow direction of the interstellar gas. The measured speeds of the dust particles were quite high, $\sim$ 15km/sec, larger than the escape velocity for the solar system at the point of detection indicating the orbits to be unbound to the solar system. The observed speeds also roughly correspond to the ones expected for interstellar grains. Supportive evidence also comes from the fact that the flux of dust particles observed were constant at all latitudes above the ecliptic plane while the interplanetary dust particles show concentration towards the ecliptic. Galileo spacecraft en route to Jupiter also encountered interstellar dust particles and gave results for mass density on interstellar dust in accord with those of Ulysses spacecraft. The mass density of interstellar dust particles measured by the two spacecrafts is shown in Fig. 3.21. It shows almost a flat distribution function. The total mass density of the observed interstellar dust is 7×10^{-27} g/cm^3.

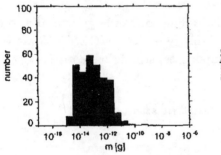 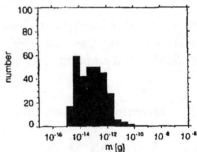

Fig. 3.21 Observed mass distributions of interstellar grains with dust instruments on Galileo (left) and Ulysses (right) spacecrafts (Grun, E., Kruger, H. and Landgraf, M. 2000, In *Minor-Bodies in the outer Solar System*, Eds. A. Fitzsimmons, D. Jewitt and R.M. West, Springer-Verlag, p.99: with kind permission of Springer Science and Business Media).

Radar studies have shown the presence of larger size particles in the range of tens of microns entering the atmosphere with speeds above 100km/sec. These speeds are well above the escape speed from the solar system arriving in a wide range of directions and is not collimated to the interstellar gas direction as smaller particles indicate. All these confirm their interstellar origin.

All the above studies carried out at the present time relate to their survey or study of their dynamical properties and mass of the dust particles. The next step will be to study the chemical composition of specific dust components and their environments. This should supplement the interstellar dust studies derived from other means.

3.13.2 Presolar Grains

Comets, Meteorites and Interplanetary Dust Particles (IDPs) provide a unique opportunity for the study of interstellar dust grains in the laboratory (Chaps. 4 and 5).

Cometary dust is believed to be an aggregate of interstellar dust particles containing carbonaceous and icy material. The dust is highly porous and fluffy in nature. The *in-situ* measurements for the chemical composition of dust particles in the coma of Comet Halley was carried out in 1986 with instruments on Giotto and Vega Spacecrafts. Several thousand mass spectra of dust particles were recorded by the instruments. These studies showed three classes of particles to be present; namely Rocky type (containing Mg, Si, Ca, Fe), CHON (containing H, C, N, O) and a mixture of these two types of particles. The core-mantle type of particles also appeared to be present. Several dust particles were found to have values of $^{12}C/^{13}C$ ratio going upto 5000. This is drastically different from the normal terrestrial value of 89. This indicates that carbon coming out of different nucleosynthesis sites has been incorporated in the Halley dust particles. Polymerized formaldehyde, $(H_2CO)_n$, also called Polyoxymethylene (POM) was seen in the dust grains of Comet Halley.

Of course the best method of studying cometary dust particles is to subject them to a thorough laboratory investigation. This is what NASA'S Stardust Mission hopes to accomplish. The Stardust Spacecraft, launched in February 1999 has already collected dust particles from Comet 81P/Wild 2 during its encounter with the comet in January 2004. The spacecraft is expected to return to Earth in 2006. The laboratory investigation of these dust particles should give a deeper insight into the elemental and isotopic ratios of cometary grains and hence its relation to grains in interstellar space. The Stardust Mission is also expected to look for interstellar grains flowing through the solar system.

Meteorites are primitive solar system materials and contains preserved nebular condensates. Therefore these primitive bodies have preserved the oldest dust material from interstellar space. Therefore several meteorites have been subjected to thorough analysis in the laboratory to look for the possible presence of these grains. These studies have shown that most primitive meteorites do contain presolar grains. The identified types of grains are given in Table 3.8.

Interplanetary Dust Particles (IDPs) collected from the Earth's atmosphere is another important source for the laboratory study of presolar

Table 3.8 Presolar grains detected in meteorites.

Composition	Diameter (μm)	Origin
C (diamond)	0.002	SN
SiC	0.3 - 20	AGB, nova
C (graphite)	1 - 20	AGB, SNII, nova
SiC (type X)	1 - 5	SN
Al_2O_3 (corundum)	0.5 - 3	RG, AGB
Si_3N_4	~ 1	SN II

Hoppe, P.M. and Zinner, E. 2000, J. Geophys. Res., **105**, 10371: Reproduced by permission of American Geophysical Union.

grains. The isotopic anomalies in several IDPs which contained relatively large refractory minerals have been seen. This is indicative of presolar material. The pyroxene whiskers appeared to be formed in the vapour phase indicating the possibility of formation in an outflow. The most promising candidate is a class of glassy silicates present in IDPs known as GEMS (Glass with Embedded Metal and Sulphides). The sizes of most of GEMS is around 0.1 to 0.5μm in diameter. This is very similar to amorphous interstellar silicate grains. The infrared absorption spectra of GEMS in the region 8-13μm is quite similar to that of interstellar silicate.

It is interesting that FeS grains have been seen from IDPs, meteorites, comets and from young and old stars. One of the IDPs in which FeS was identified was rich in deuterium, indicating the source to be from a molecular cloud. Therefore the FeS grain from this IDP is presolar in origin. The presence of FeS grains in all kinds of objects mentioned above give a direct link between the solar system material to the presolar material. It was seen earlier that in the diffuse interstellar medium sulphur is not depleted much (Fig. 3.11). But the situation is vastly different in molecular clouds where sulphur appears to be highly depleted in the gas phase. Therefore it is likely that this sulphur is tied up in grains as FeS. Hence it is likely that the solar system bodies formed out of the solar nebular material had already crystalline FeS grains.

In conclusion, all the observations existing at the present time show that silicate is the dominant component of interstellar dust. It is largely amorphous with crystalline fraction of atmost 2%. However the exact composition of interstellar silicate is rather uncertain. The carbonaceous material form the other component of interstellar grains. They include dust grains such as diamond, graphite, amorphous or glassy carbon, PAH etc. Some of these identifications have come from the study of presolar grains in comets, meteorites and IDPs. The dust grains collected by Stardust mission during its encounter with Comet 81P/Wild 2 in January 2004, when subjected to laboratory investigation after the return of the spacecraft in 2006, should give a better understanding of the chemical composition of dust in interstellar space. The stardust mission is also expected to collect interstellar dust grains passing through the solar system. The laboratory studies of these grains will provide chemical composition of these particles directly. Many emission nebulae show Unidentified Infrared Emission features (AUIBs). The dust grains present in the sources which are embedded in a heavily obscured material has a highly complex spectra with several strong absorption features such as CO, CO_2, CH_4 etc. These features arise out of volatile mantles condensed on refractory dust cores. The extinction law towards galactic center deviates from the standard graphite-silicate corresponding to the general interstellar extinction curve. Bulk of the particles present in interstellar space originate in oxygen-rich M giants producing dust rich in silcon, metals and oxygen and carbon-rich stars producing carbonaceous type of particles.

References

The following publications are very useful
Dorschner, J. 2001, In *Interplanetary Dust*, Eds. E. Grun, Bo. A.S. Gustafson, S. Dermott and H. Fechtig, Springer-Verlag, p.727.
Drain, B.T. 2003, Ann. Rev. Astron. Astrophys., **41**, 241.
Greenberg, J.M. 1978, In *Cosmic Dust*, Ed. J.A.M. McDonnell, John Wiley and sons, p.187.
Greenberg, J.M. and Shen, C. 1999, Astrophys. Space Science, **269**, 33.
Hoyle, F. and Wickramasinghe, N.C. 1991, *The Theory of Cosmic Grains*, Kluwer Academic Publishers, Dordrecht.
Krugel, E. 2003, *The Physics of Interstellar Dust*, Institute of Physics Publishing, Bristol.
Whittet, D.C.B. 2003, *Dust in the Galactic Environment*, Institute of Physics Publishing, Bristol.

PAHs and Interstellar Extinction curve
Donn, B. 1968, Astrophys. J., **152**, L129.
Donn, B. and Krishna Swamy, K.S. 1969, Physica, **41**, 144.
Duley, W.W. and Seahra, S. 1998, Astrophys. J., **507**, 874.
Platt, J.R. 1956, Astrophys. J., **123**, 486.
Zubko, V., Dwek, E. and Arendt, R.G. 2004, Astrophys. J. Suppl., **152**, 211.

Extinction, Size distribution
Krishna Swamy, K.S. 1965. Publ. Astron. Soc. Pacific, **77**, 164.
Mathis, J.S. 1990, Ann. Rev. Astron. Astrophys., **28**, 37.
Mathis, J.S., Rumpl. W. and Nordsieck, K.H. 1977, Astrophys. J., **217**, 425 (MRN).

Detection of 2175Å feature and graphite model proposal
Hoyle, F. and Wickramasinghe, N.C. 1962, Mon. Not. Roy. Astron. Soc., **124**, 417.
Stecher, T.P. 1965, Astrophys. J., **142**, 1683.
Stecher, T.P. and Donn, B. 1965, Astrophys. J., **142**, 1681.

Early work on Diffuse Galactic Light is from
Henyey, L.G. and Greenstein, J.L. 1941, Astrophys. J., **93**, 70.

For later work
Lillie, C.F. and Witt, A.N. 1976, Astrophys. J., **208**, 64.
Morgan, D.H., Nandy,K. and Thomson, G.I. 1978, Mon. Not. Roy. Astron. Soc., **185**, 371.
Witt, A.N. 1968. Astrophys. J., **152**, 59.
Witt, A.N. 1989, In *Interstellar Dust*, Eds. L.J. Allamondala and A.G.G.W. Tielens, Kluwer Academic Publishers, p.87.

Scattering from Reflection Nebulae, HII Regions
O'Dell, C.R. and Hubbard, W.B. 1965, Astrophys. J., **142**, 591.
Smith, T.L. and Witt, A.N. 2002, Astrophys. J., **565**, 304.
Witt, A.N., Peterson, J.K., Holberg, J.B., Murthy, J., Dring, A. and Henry, R.C. 1993., Astrophys. J., **410**, 714.
Witt, A.N. and Schild, R.E. 1986, Astrophys. J. Suppl., **62**, 839.

Extended Red Emission
Duley, W.W. 2001, Astrophys. J., **553**, 575.
Ledoux. G., Ehbrecht, M., Guillois, O., Huisken, F., Kohn, B., Laguna, M.A., Nenner, I., Vaillard, V., Papoular, R., Porterat, D. and Reynaud, C. 1998, Astron. Astrophys., **333**, L39.
Seahra, S. and Duley, W.W. 1999, Astrophys. J., **520**, 719.
Schmidt, G.D., Cohen, M and Morgan, B. 1980, Astrophys. J., **239**, L133.
Watanabe, I., Hasegawa, S. and Kurata, Y. 1982, Japanese J. Appl. Phys. **21**, 856.
Witt, A.N., Gordon, K.D. and Furton, D.G. 1998, Astrophys. J., **501**, L111.

Early work on Elemental Depletion
Field, B. 1974, Astrophys. J., **187**, 453.
Morton, D.C., Drake, J.F., Jenkins, E.B., Rogerson, J.B., Spitzer, L. and York. D.G. 1974, Astrophys. J., **181**, L103.

For later work
Jenkins, E.B. 1989, In *Interstellar Dust*, Eds. L.J. Allamondala and A.G.G.M. Tielens, Kluwer Academic Publishers, p.23.
Mayer, D.M., Jura, M. and Cardelli, J.A. 1998, Astrophys. J., **493**, 222.
Sophia, U.J. 1997, In *Stardust to Planetesimals*, Eds. Y.J. Pendleton and A.G.G.M. Tielens, ASP Conference Series, Vol.122, p.77.
Sophia, U.J., Cardelli. J.A. and Savage, B.D. 1994, Astrophys. J., **430**, 650.

Diffuse Interstellar Bands
Herbig, G. 1995, Ann. Rev. Astron. Astrophys., **33**, 19.
Ruiterkamp. R., Halasinski, T., Salama, F., Foing, B.H., Allamandola, L.J., Schmidt, W. and Ehrenfreund, P. 2002, Astron. Astrophys., **390**, 1153.
Tielens, A.G.G.M. and Snow, T.P. (Eds.) 1995, *The Diffuse Interstellar Bands*, Kluwer Academic Publishers.

For Infrared Spectral Features, see the excellent publication
d'Hendecourt, L., Joblin, C. and Jones, A. (Eds.) 1999, *Solid Interstellar Matter: ISO Revolution*, Springer-Verlag.

Infrared Emission.
Diffuse Interstellar Medium
Pendleton, Y.J. and Allamandola, L.J. 2002, Astrophys. J. Suppl., **138**, 76.
Whittet, D.C.W. and Tielens, A.G.G.M. 1997, In *From Stardust to Planetesimals*, Eds. Y.J. Pendleton and A.G.G.M. Tielens, ASP Conference Series, Vol.122, p.161.

HII Regions
Cesarsky, D., Jones, A.P., Lequeux, X. and Verstraete, L. 2000, Astron. Astrophys., **358**, 708.
Stein, W.A. and Gillett, F.C. 1969, Astrophys. J., **155**, L197.

Reflection Nebulae
Moutou, C., Sellgren, K., Leger, A., Verstraete, L., Rouan, D., Giards, M. and Werner, M. 1998, In *Star Formation with Infrared Space Observatory*, ASP Conference Series, Vol.132,Eds. J.L. Yun and R. Liseau, p.47.
Sellgren, K., Allamandola, L.J., Bregman, J.D., Werner, M.W. and Wooden, D.H. 1985, Astrophys. J., **299**, 416.

$FeSi_2$
Gordon, K.D., Witt, A.N., Rudy, R.J., Dutter, P.C., Lynch, D.K., Mazuk, S., Misselt, K.A., Clayton, G.C. and Smith, T.L. 2000, Astrophys. J., **544**, 859.

Molecular Clouds
Gibb, E.L., Whittet, D.C.B., Boogert, A.C.A. and Tielens. A.G.G.M. 2004, Astrophys. J. Suppl., **151**, 35.

Galactic Center

Chair, J.E., Tielens, A.G.G.M., Whittet, D.C.W., Schutte, W., Boogert, A., Lutz, D., van Dishoeck, E.F. and Bernstein, M.P. 2000, Astrophys. J., **537**, 749.

Dwek, E. 1998, Astrophys. J., **501**, 643.

Gibb, E.L., Whittet, D.C.B., Boogert, A.C.A. and Tielens, A.G.G.M. 2004, Astrophys. J. Suppl., **151**, 35.

Lutz, D., Feuchtgruber, H., Genzel, R., Kunze, D., Rigopoulou, D., Spoon, H.W.W., Wright, C.M., Egami, E., Katterloher, R., Sturm, E., Wieprecht. E., Sternberg, A., Moorwood, A.F.M. and Graauw, Th.de. 1996, Astron. Astrophys., **315**, L269.

Roche, P.F. 1988, In *Dust in The Universe*, Eds. M.E. Bailey and D.A. Williams, Cambridge University Press, Cambridge, p.415.

Sources of Dust

Gehrz, R.D. 1989, In *Interstellar Dust*, IAU Symposium No.135, Eds. L.J Allamandola and A.G.G.M. Tielens, Kluwer Academic Publishers, p.445.

Jones, A.P. 1997, In *Stardust to Planetesimals*, Eds.Y.J. Pendleton and A.G.G.M. Tielens, ASP Conference Series, Vol.122, p.97.

Whittet, D.C.B. 2003, *Dust in the Galactic Environment*, Institute of Physics Publishing, Bristol, p.257.

in-situ detection of dust

Grun, E. et al., 2001, In *Interplanetary Dust*, Eds. E. Grun, Bo. A.S. Gustafson, S. Dermott and H. Fechtig, Springer-Verlag, p.295.

Meisel, D.D., Janches, D. and Mathews, J.D. 2002, Astrophys. J., **567**, 323.

Chapter 4

Cometary Dust

4.1 Introduction

The various objects comprising the solar system were formed from the solar nebular material roughly at the same time. But planets being massive objects have undergone various physical processes since its formation that has resulted in their original material to be modified considerably. But comets being very small and cold objects not much has happened since the time of their formation. Hence comets are believed to have preserved their original solar nebular material composition intact. Therefore comets are the most pristine bodies in the solar system. Hence the study of dust of comets is important to understand the origin of the solar system from the interstellar material.

When the comet is far off from the sun it looks as point source of light. This is just the reflected sunlight from the nucleus of the comet the dimension of which is around 1 to 10km. As it approaches the sun the volatiles and the dust comprising the nucleus of a comet sublime due to solar heating. The gas and the dust expand radially resulting in the cloud around the nucleus called the Coma. The coma becomes brighter and larger in size with the decrease in sun-comet distance. The coma is made visible through emission lines from molecules and by scattered radiation by dust particles. The dimensions of the visible coma is around 10^4 to 10^5km while the ultraviolet coma resulting from Lyman α emission extends upto about 10^6 to 10^7km. The spectacular feature of a comet is the presence of two tails the dust tail and the plasma tail. The extent of the two tails is maximum close to the perihelion distance. The dust released by the nucleus is dragged outwards by the gas accompanying it and at the same time the dust is acted upon by the radiation pressure of the sun which pushes them away from the sun. The resultant effect is to give a curvature to the dust particles which gives rise to the curved dust tail. The dust tail extends upto about 10^7km or so. The dust tail is made visible through scattered solar radiation by the submicron size dust particles in the tail. In contrast to the normal dust tail which point to the direction away from the sun quite often a short tail is seen in the direction towards the sun. This is called the anti-tail of the comet which contain larger size particles (few microns) than those present in the normal dust tail. Dust trails have also been seen which are narrow trails of dust of centimeter in size and located along the orbit of the comet (Sec. 4.2.2). The particles in the anti-tail are ejected with velocities close to zero and lie in the comet orbital plane. When the orbit of the comet crosses the orbital plane of the Earth it is seen as a narrow tail pointing towards the sun. The gas in the coma comprising mostly molecules is subjected to solar radiation which ionizes them. These ions are swept outwards radially by the stream of charged particles present in the solar wind. Solar Wind is basically the solar plasma from the corona that flows out continuously into the interplanetary medium ($\sim$430km/sec at 1AU). These two are coupled through the interplanetary magnetic field. This gives rise to the plasma tail or ion tail and extends upto about 10^7 to 10^8km. The spectrum of the plasma tail shows mainly emission from ions like CO^+, CO_2^+, N_2^+ etc; of these the emission due to CO^+ is the dominant one. The dust tail is generally very smooth and structureless while the plasma tail contain large scale structures of various kinds.

The nucleus of a comet believed to contain the unaltered primitive material of the solar system is hard to study directly because of its small size. In addition the nucleus is hidden by the cloud of gas and dust surrounding it. Therefore the nature and chemical composition of the nucleus of a comet has to come from indirect means basically from the study of the gas and dust in the coma. A reasonable working model for the nucleus was first proposed by Whipple in the 1950's. This is generally called icy-conglomerate model. In this model the nucleus is believed to be a discrete rotating body consisting of frozen water, complex molecules formed out of abundant elements H, C, N and O, and dust. All the subsequent observations of comets for the last four decades have supported this model by indirect means. This basically came through the observation of the dissociated products of H_2O, namely OH, H and O, and H_2O^+. The first direct observation of H_2O in a comet came with the detection of 2.7 μm band of H_2O in Comet Halley. The nucleus contains around 80% of H_2O-ice and the rest is made up of other constituents and dust. The solid body nature of the nucleus in contrast to that of 'loosely bound system' was confirmed by photographs taken of the nucleus of Comet Halley by Giotto spacecraft when it was at a distance of 600km from the nucleus supporting the general concept of Whipple's model.

As can be seen from the above brief discussion dust forms an important component in the nuclei of comets. The understanding of the nature, composition and sizes of the dust particles in comets comes mainly from the study of the dynamics of dust tails and from the interpretation of continuum, polarization and infrared observations with their associated spectral features. Except for the infrared observations of comets, other studies provide an average property for the dust particles. Of course the best way for the study of dust particles is by making *in-situ* measurements of various kinds in the coma of comets. This was actually carried out for the first time by six spacecrafts belonging to various space agencies during March 1986, around the time of the closest approach of Comet Halley to the Sun (Fig. 4.1). The closest approach to the nucleus was made by the European Space Agency's spacecraft Giotto, which passed at a distance of approximately 600km from the nucleus. The spacecrafts to Comet Halley carried variety of scientific experiments and performed a wide range of *in-situ* measurements. Some of the experiments complemented each other while others overlapped. Due to the retrograde motion of Comet Halley with respect to the Earth and the spacecraft, the relative encounter velocity was very high $\sim$ 78km/sec. These missions were highly successful and have increased our understanding on the nature of dust particles in comets in a dramatic way. Some of these aspects will be covered in this chapter.

There are various indications that the cometary dust is porous, fluffy and made up of agglemeration of smaller sized particles. The cometary dust is basically the aggregation of interstellar dust particles containing silicates, ices and refractories. Interplanetary dust particles which are believed to be coming from comets, appears as aggregates of large number of nearly spherical particles of about $0.1\mu m$ in size. The existence of porous dust in the coma is postulated inorder to explain the infrared emissions. The ice and other carbonaceous volatiles trapped in the porous aggregates evaporates due to solar heating. Therefore the dust particles in the coma of comets may be defined as "aggregate submicrometer grains" that have been sintered into larger particles using ice and organic materials. This is necessary to make the distinction between the interstellar grains and the cometary grains and interplanetary dust particles.

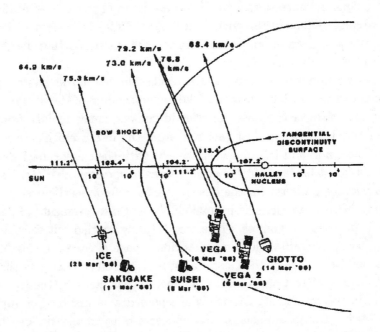

Fig. 4.1 The geometry of the six spacecraft flybys to Comet Halley. The distances are marked in logarithmic scale and the Sun is to the left of the comet. The flyby dates for each mission are given at the bottom, flyby phase angle in the center and flyby speeds at the top (Mendis, D.A. Reprinted with permission from Annual Review of Astronomy and Astrophysics, volume 26 (c) 1988 by Annual Reviews www.annualreviews.org).

4.2 Dust Tails

4.2.1 *Dynamics of Dust Tails*

The dust released by the nucleus is acted upon by two opposite forces. The solar radiation pressure acting on the dust particles away from the sun while the solar gravitational attraction tries to pull it towards the sun. Since the two forces vary as $1/r^2$, it is convenient to describe an effective gravity essentially given by the difference of the two forces. The Keplerian orbit mechanics is then used for the study of the dynamics of the dust particles. The ratio of radiation pressure to gravitational force is generally denoted by $(1-\mu)$ where

$$\beta \equiv (1-\mu) = \frac{F_{rad}}{F_{grav}}$$
$$= 1.2 \times 10^{-4}\frac{Q_{pr}}{\rho d}. \qquad (4.1)$$

Here ρ is the density and d is the diameter of the particle. Q_{pr} is the efficiency factor for radiation pressure and depends upon the material property of the dust and its size. The nature of the orbit of the dust particles is decided by the value of β. The complete description of the dynamics of dust tails was formulated by Finson and Probstein in 1968. There are two cases of interest that have to be considered. Firstly the particles are ejected continuously as a function of time by the nucleus. The locus of the particles which have the same value of β is called Syndyne curve. Physically this means that particles of the same size are emitted by the nucleus. The other case involves the distribution of particles emitted at any one particular time as in the case of an outburst. These particles will have varying β values and the locus of the curve which describes this situation is called Synchrone. Some of the calculated Syndynes and Synchrones are shown in Fig. 4.2. The further one goes along the tail, the earlier the time at which the par-

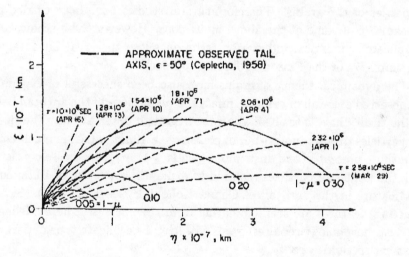

Fig. 4.2 Syndynes and Synchrones for Comet Arend-Roland on April 27, 1957. Time $\tau = 1.71 \times 10^6$ sec at perihelion on April 8. The position of the observed tail of the comet is also shown (Finson, M.L. and Probstein, R.F. 1968, Astrophys. J., **154**, 327 and 353: reproduced by permission of the AAS).

ticles were emitted. However in a real situation the particles are emitted with a certain velocity spread which gives rise to the observed spread in the tail. Therefore dust particles in the tail contains particles of various sizes emitted by the nucleus at various earlier times. The Finson-Probstein model sums over the contributions from these two sources. From a comparison of the expected and the observed intensity distributions it is possible to make an estimate of the production rate of grains, velocity of the grains and the size distribution function. The exponent α in the size distribution function of the form n(a)$\propto a^{-\alpha}$ (a is the size of the grain) derived from the study of several comets, give values $\sim$ 3.2 to 3.7. This is consistent with the value of $\alpha \sim$ 3.5 deduced from a fit to infrared observations of comets (Sec. 4.7) and also from the study of interstellar extinction curve (Sec. 3.5.2). Even though Finson and Probstein formalism has been successful in the studies of tail morphology there are several limitations. The grains were assumed to be of the same kind. However both dielectric and absorbing grains are present in comets with a wide variation in their sizes. The efficiency factor for radiation pressure, Q_{pr}, was assumed to be unity. i.e. independent of the composition of the dust particles. However in a real situation it depends on size and composition of the dust particles. If the grains of various sizes are emitted continuously it is difficult to separate the effects depending on the time of ejection from those due to the size and properties of the grains. Therefore several studies have been carried out to take care of some of the above limitations. However these studies with complicated numerical integration techniques lose the great advantage of the simplicity of the Finson-Probstein approach.

The dynamical theory of grains have also been successful in explaining the observed anti-tail of comets. Infact the predictions of the anti-tail based on the theory have later been confirmed through observations. The size of the particles present in anti-tail of comets is $\geq 10\mu$m, which is much larger than those present in the dust tails which is $\leq 1\mu$m. This is also consistent with the large size particles required to explain the absence of 10μm emission feature in the anti-tail of Comet Kohoutek (see Fig. 4.14). Several structural features are seen eminating from the nucleus. The modeling of the dust morphology requires grains with $\beta < 1$ i.e silicate type of particles (see next section).

4.2.2 Dust Trail

Dust Trails were first seen in the IRAS observations of Comet Temple 2, which are extremely narrow with the length-to-width ratio of approximately 200 to 1. Some trails were found to coincide with the projection on the sky of comet orbits. Dynamical calculations indicate that the observed dust trails are coincident with large particle trajectory indicating the presence of large particles of millimeter to centimeter in size. These particles are ejected at low velocities $\sim$a few m/s from the nucleus and are least affected by the solar radiation pressure. The emission from the dust must have occurred at least 1500 days (about 1 orbital period) prior to the IRAS observations. The persistence of the narrowness of the trail indicate the velocity dispersion to be very low, <2m/s. Therefore these trails appear to be stable against perturbing forces over periods of atleast a few orbital revolutions. Hence the dust trails is the resultant effect of the superposition of large particle emissions over several orbital periods. The ISO observations could spatially resolve the dust trail in Comet P/Encke. The detection of dust trails has come from the observations carried out with spacecrafts in the infrared wavelengths. It has been the feeling that the dust trails are too faint to be seen with ground based observations in the visible region (because of very large size particles). However it has been detected in the visible region(λ_{eff}=0.65μm) along the orbit of Comet 22P/Kopff, when the comet was at r=3.01AU. Therefore the study of dust trails of comets in the visible region should help in the understanding of the nature of very large size dust particles. Therefore dust trails are in contrast to the bright and broad dust tails of comets which are composed of submicron size dust particles pushed out by the effect of solar radiation pressure.

4.3 Radiation Pressure Effects

As mentioned earlier the ratio of solar radiation pressure to gravity on the dust particles is expressed by the parameter $\beta \equiv (1-\mu)$. From the photographs of the dust tail and from dynamical considerations it is possible to estimate the range of β values which encompasses the observed tail. Other particles will essentially be pushed out of the system. From such comparisons it is possible to estimate the maximum value of β denoted as β_{max} for various cometary tails. Observations indicate β_{max}=2.5. This apparent cut off in the values of β can be used to infer the properties of the dust particles. The interpretation invokes various models incorporating size, shape and optical properties of the grains. Most of the studies invoke Mie Theory of scattering which is for compact spherical grains of particular composition. As an example some of the results of such calculations is shown in Fig. 4.3. As can be seen from this figure, absorbing grains have a peak in β greater than unity at particle sizes of near 0.1μm and the curve becomes flatter for very small size particles with a nonzero β value. On the other hand dielectric particles have a peak in β around particle sizes of 0.2 to 0.3μm and falls down steeply for very small sizes, with $\beta \sim 0$.

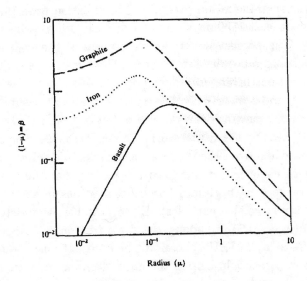

Fig. 4.3 The variation of radiation pressure force $(1-\mu) \equiv \beta$ with the radius of the particles for three types of material is shown (Adapted from Saito, K., Isobe, S., Nishioka, K. and Ishii, T. 1981, Icarus, **47**, 351).

4.4 Visible Continuum

Comets in general possess continuua in the visible region of the spectrum due to scattering of the solar radiation by submicron dust particles. The strength of the continuum depends upon the dust content. Therefore dusty comets should have stronger continuum. Since the scattering of radiation by dust particles is a function of wavelength and scattering angle, the intensity of the scattered radiation depend upon the nature and composition of dust grains. Therefore in principle it is possible to infer the properties of the dust particles from a study of the continuum. But the continuum is affected by the presence of strong molecular lines due to C_2, CN, CH and NH_2. Therefore the continuum has to be corrected for these emission features or a spectral region has to be selected where the emission features are absent or minimal. From these observations it is possible to determine the relative continuum intensities of comet with respect to the Sun or a star of similar to the Sun. This yields the wavelength dependence of the observed continuum of the comet. By combining the data obtained in the ultraviolet region with the ground-based observations it has been possible to get the continuum curve for the coma in the wavelength region 0.26 to $3\mu m$. This showed the reddening to be significant in the ultraviolet region and decrease slowly with an increase in wavelength.

The interpretation of the observed reddening should give information about the nature of the dust particles. However it is complicated as it involves several parameters. In general, the scattered radiation is a function of wavelength of the incident radiation (λ), scattering angle (θ), the refractive index of the material (μ=n+ik) and the size distribution of the particles(a). i.e I (a,λ,μ,θ). The mean scattering intensities can be calculated from the Mie Theory for spherical particles or from other theories for non spherical particles for a given material property for comparing with the observed variations.

The early continuum observations of the tails of Comets Arend-Roland and Mrkos indicated that grains with refractive index of iron and sizes $\sim$ 0.3 μm are most appropriate. Later, studies of other comets indicated the particle sizes to be submicron to micron and slightly absorbing. Extensive calculations have been carried out for various types of silicate materials and for constant values for the refractive index(n) with a small absorption part(k) for restricting the ranges in n and k which fit the observations. These comparisons indicate n$\approx$ 1.5 to 1.6, k$\leq$ 0.05 and the particle size $\sim$ 0.2μm. The values of n roughly corresponds to those of silicate type of material.

4.5 Phase Function

The scattering intensity as a function of scattering angle i.e. phase function can also be used for extracting material grain property. The phase function can be obtained by observing the coma of the comet as a function of the heliocentric distance or by scanning along the outward direction of the tail. In such observations the quantity that is changing is basically the scattering angle. It is considerably difficult to derive the phase function since the physical conditions within the coma change with the scattering angle. The aproximate phase function has been derived from the ratio of the visual flux which represents the scattered radiation to the infrared brightness that represents the absorbed energy for the scattering angles between 30° and 150°. Since they both come from the same volume of dust the ratio is simply proportional to the scattering function of the grains. The derived phase function for several comets is shown in Fig. 4.4a. The phase function shows a forward scattering lobe for scattering angle $\leq$ 40°

with almost a flat shape for the angle between 60° and 150°. The nature of the observations is more in conformity with the expected results from non-spherical particles (Fig. 4.4b). The observed phase function is consistent with the expected distribution of micron size particles with an index of refraction in the range $1.3 \leq n \leq 2.0$ and $k \geq 0.05$.

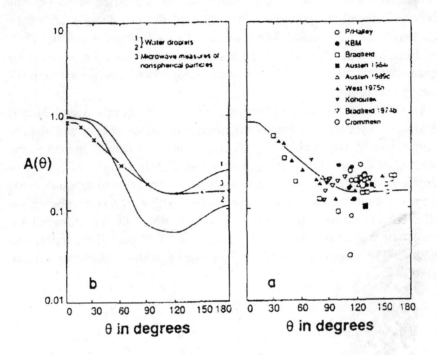

Fig. 4.4 (a) A composite phase diagram of the observed ratio of the reflected to the infrared flux as a function of the scattering angle derived for several comets. For comparison with the data, the curve 3 from (b) is also shown. (b) shows the laboratory data from spherical (curves 1 and 2) and non-spherical particles (curve 3) (Gehrz, R.D. and Ney, E.P. 1992, Icarus, **100**, 162).

4.6 Polarization

The polarization of the scattered radiation is another important observation which can give information on the nature and the size of the particles. Infact the polarization is quite sensitive to the shape, structure and sizes of the dust particles. The variation of polarization as a function of scattering angle θ has been established from the observations of a number of comets.

There is some variation in the observed polarization with scattering angle θ. In particular, the polarization values tend to divide into two groups with $P_{max} \sim 10-15\%$ and $\sim 25\%$ respectively for smaller scattering angles. The average behaviour of the polarization curve as a function of the scattering angle is shown in Fig. 4.5. The polarization shows negative values $\sim -2\%$ for scattering angle $\geq 160°$, a gradual increase with the decrease in scattering angle reaching maximum polarization value ~ 15 to 25% around $\theta \sim 90°$ and then decrease for θ's $< 90°$. The extensive polarization measurements carried out in Comet Halley from the visible to near-infrared wavelength regions show a very similar behaviour as shown in Fig. 4.5. This general behaviour could be taken as a representative of quiet conditions of the coma. There could be variation of polarization with the level of activity of the nucleus such as the presence of jets, bursts or in active comets.

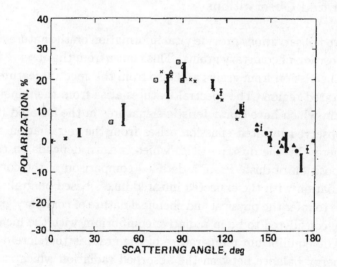

Fig. 4.5 The observed dust polarization in several comets denoted by various symbols is plotted as a function of the scattering angle. The existence of negative polarization for scattering angles around 170° can be seen. The observations clearly delineate a well defined polarization curve as a function of the scattering angle. The dark vertical bars are the calculated values (Krishna Swamy, K.S. and Shah, G.A. 1988, Mon. Not. Roy. Astron. Soc., **233**, 573).

For interpreting polarization observations, Mie theory of scattering for spherical particles has been used extensively. From a comparison of the observed polarization with those of calculations it is possible to restrict the ranges in the input parameters for the dust particles. The usual procedure adopted is to calculate the polarization as a function of the scattering angle for a grid of real and imaginary part of the refractive indices, particle sizes and then try to get the best fit to the observations. Such a procedure gave limit on n and k of $1.3 < n < 2.0$ and $0.01 < k < 0.1$ and particle size $\geq 1\mu m$. The analysis of extensive polarization measurements carried out on Comet Halley gave the average refractive index of the grain for the visible region, $n=1.39\pm0.01$ and $k=0.035\pm0.004$. In this study the size distribution of grains in Comet Halley derived from Vega measurement was used. The observed linear polarization with scattering angle for several comets is also consistent with such a refractive index. The calculations for rough spheres including both silicate and graphite grains also gave a good fit to the Comet Halley polarizaton measurements.

4.7 Infrared Observations

The infrared observations provide vital information on the nature and chemical composition of cometary grains. This comes from the observed strength of thermal emission from grains and also from the spectral features present in the infrared region. The spectral features arise from the mineral phases of the grain which have characteristic signatures in the infrared (Sec. 2.2).

The observed infrared emission arises from the re-radiation of the absorbed energy by the dust particles which in turn depends on the nature and composition of dust. From a detailed comparison of the cometary infrared radiation with the expected infrared fluxes based on grain models it is possible to infer the physical and chemical nature of cometary grains. The grains are considered to be in radiative equilibrium with the incident solar energy. The equilibrium temperature of the grain is therefore determined by the energy balance between the absorbed radiation which is mostly in the ultraviolet and visible regions and the emitted radiation which is in the far-infrared region. This can be expressed by the condition

$$F_{abs}(a) = F_{em}(a, Tg) \tag{4.2}$$

where

$$F_{abs}(a) = (R_\odot/r)^2 = \int F_\odot(\lambda) Q_{abs}(a, \lambda) \pi a^2 d\lambda \tag{4.3}$$

and

$$F_{em}(a, Tg) = \int \pi B(\lambda, Tg) Q_{abs}(a, \lambda) 4\pi a^2 d\lambda \tag{4.4}$$

where $F_\odot(\lambda)$ is the incident solar radiation at wavelength λ, $Q_{abs}(a,\lambda)$ is the absorption efficiency of the dust particles and $B(\lambda,Tg)$ is the Planck function corresponding to the grain temperature Tg. The difference in the geometric areas between the two sides arises as the grains absorb in one direction and emit in all directions. The calculation of grain temperature involves a knowledge of the grain property and its size. Knowing the grain temperature it is possible to calculate the emission from the dust particles. In general the emission has to be integrated over the size distribution function and also sum over various types of grains with assumed grain properties that could be present in the coma, to get the total thermal emission from the dust particles.

The first detection of excess infrared radiation in the wavelength region from 1 to 10μm was from Comet Ikeya-Seki in 1965. It showed that the colour temperature was higher than the black body temperature at the same heliocentric distance. These results have been confirmed based on observations of other comets. Therefore infrared excess observed in comets is a common property of all comets. Most of the observations on comets before Comet Halley were limited to broadband infrared observations in the spectral region around 2 to 20μm (Fig. 4.6). However for Comet Halley it has been possible to get very good broadband and spectroscopic data in the mid-infrared and far-infrared wavelength regions based on ground-based, airborne and spaceborne instrumentation. So there exists infrared data in the wavelength region from around 3 to 160μm.

The infrared observations of earlier Comets Kohoutek, Bradfield and Bennett could be fitted with dust particles of the type n=1.6 or 1.33 and with k $\approx$ 0.05. However in recent times attempts have been made to reproduce the observed infrared radiation from comets by taking into account the variation of optical properties for cometary dust particles in the calculation of thermal emission from the dust grains. These models are based on simple grain population or a mixture of several types. The two major components to the grain material are the silicate rich olivine and some form of carbon (Sec. 4.9). The use of the above components for the grain material and the size distribution of the form n(a) $\propto a^{-\alpha}$ with $\alpha \sim 3.5$, it is possible to explain the broadband infrared observations of several comets. The above size distribution function is consistent with the size distribution derived from the study of dynamics of dust tails and also similar to the size distribution of interstellar grains.

The production rate of grains can be calculated from a knowledge of the total number of dust particles in the field of view to explain the observed

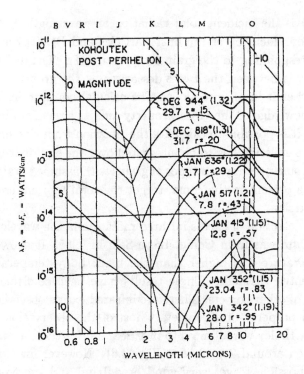

Fig. 4.6 Observed infrared fluxes plotted as a function of wavelength for Comet Kohoutek. The fitted Black body temperature for each of the curves along with the factor by which the temperature exceeds the black body temperature is also shown (Ney, E.P. 1974, Icarus, **23**, 551).

infrared flux at the Earth. The value derived for Comet Halley is $\approx 4 \times 10^6$ kg/s at 1AU. The production rate of grains varies from comet to comet. Dusty comets will have larger dust production rate compared to less dusty comets. It also varies with the heliocentric distance for the same comet.

4.8 Albedo

The average value for the albedo of cometary particles can be obtained from a combination of simple physical arguments and observations. The observation required for such a study is the scattered radiation at the visible wavelength and the total integrated infrared flux of the comet. The basic physical idea behind such a simple method is that the continuum in the visible region arises due to the scattering process which depend on the scattering efficiency of the grain. On the other hand the observed infrared radiation is dependent on the amount of impinging energy absorbed by the grain, which depends on the absorption efficiency of the grain. For small optical depths and also neglecting phase dependence effects, the optical surface bightness can be written as

$$S_{opt}(\lambda) = \frac{F_\odot(\lambda)\tau}{4\pi} = \frac{F_\odot(\lambda)N_d l \pi a^2 Q_{sca}(a,\lambda)}{4\pi}. \qquad (4.5)$$

Here N_d is the column density of the dust particles and l is the representative path length. The infrared surface brightness can be approximated as

$$S_{ir} = \frac{\langle Q_{abs}\rangle \pi a^2 F_\odot N_d l}{4\pi}. \qquad (4.6)$$

Here $\langle Q_{abs}\rangle$ represents a mean absorption efficiency for the grain and $F_\odot = \int F_\odot(\lambda)\,d\lambda$, the integrated solar radiation. It is possible to derive an expression for the average albedo γ of the particles from the above equations. This is given by

$$\frac{\gamma}{1-\gamma} = \frac{F_\odot S_{opt}(\lambda)}{S_{ir} F_\odot(\lambda)}. \qquad (4.7)$$

The estimated values of albedo for several comets lie typically in the range of about 0.1 to 0.3. Allowing for the phase dependent scattering, the value of albedo for Comet West lies in the range of about 0.3 to 0.5. Therfore the derived albedo is low and indicate refractory type of particles. The spatial variation of an average albedo of dust grains in Comet Halley at r=1.7AU has been determined from the above method and the value lies in the range of around 0.25 to 0.45.

Another form of albedo that has been used frequently is called the geometric albedo of the particle. The geometric albedo A_p, is defined as the ratio of the energy scattered by the particle at scattering angle $\theta=180°$ (backward scattering) to that scattered by a white Lambert disk of the same geometric crosssection. It is convenient to define the geometric albedo at a given scattering angle θ as $A_p(\theta)$, which is just the product of the geometric albedo and the normalized scattering function. The geometric albedo $A_p(\theta)$ has been determined for several comets from the simultaneous measurements of the scattered and thermal radiation in the J(1.25μm) and K(2.2μm) bands. The derived albedo in the J band pass varies from 0.025 for small scattering angles to 0.05 near $\theta \sim 180°$. Studies carried out in the laboratory on irregular particles with absorption properties give similar values, $A_p \sim 0.08$ for particles of radius 2μm and $\sim$0.025 for particle radius of 60μm. Models based on porous aggregates also seem to give similar trend with the variation in the size of the particle.

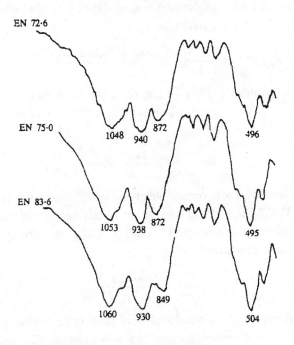

Fig. 4.7 Absorption spectra of orthopyroxene ionosilicates. Sample curves shown (in cm^{-1}) are for the enstatite-hypersthene series of En_{73}, En_{75} and En_{84} (Adapted from R.J.P. Lyon. 1963, *NASA Technical Note*, NASA TN D-1871).

4.9 Spectral Features in the Infrared

4.9.1 *Spectral Features*

The nature of grain can be inferred from the presence of specific spectral features in the infrared region. Laboratory measurements of the absorption spectra of various types of silicate material carried out at room temperature or at liquid nitrogen temperature have a common property in showing a strong and broad band around 10 μm. This is attributed to Si-O stretching vibrations (Fig. 4.7). For the first time, a broad emission feature centered around 10μm was detected in Comet Bennett (Fig. 4.8). From the similarity between the laboratory spectra and the cometary spectra it is suggested that this 10 μm feature observed in Comet Bennett is due to some type of silicate material. The other feature present in the laboratory spectra around 20 micron, arising out of Si-O-Si bending vibrations has also been seen in cometary spectra confirming the silicate nature of dust.

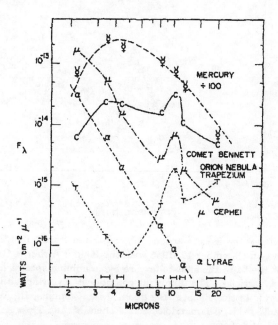

Fig. 4.8 The 10μm feature observed in Comet Bennett (Mass, R., Ney, E.P. and Woolf, N.J. 1970, Astrophys. J., **160**, L101: reproduced by permission of the AAS).

The laboratory spectra also show several other secondary features which are characteristics of the particular type of silicate material. It is quite possible that they could be washed out due to the superposition of a variety of possible types of silicate material that could exist in the coma. However if these features show up it could be used as a diagnostic of the dust composition. The observation carried out on Comet Halley with Kuiper Airborne Observatory showed clearly a well defined spectral feature at 11.2 μm located inside the broad 10μm emission feature (Fig. 4.9). This feature is attributed to crystalline olivine[$(Mg,Fe)_2SiO_4$] based on good spectral match with the observed spectral emissivity of Mg-rich olivine. The strong 10μm emission feature has been observed from a number of comets. Comet Hale-Bopp had the strongest silicate emission among all the observed comets. It was also unusual in showing a strong silicate feature even at 4.6AU preperihelion. A typical 10μm spectral feature observed in comets is shown in

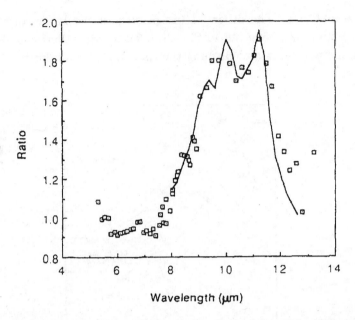

Fig. 4.9 The profile of the feature around 10μm derived from 5 to 13μm spectrum of Comet Halley taken on December 12, 1985 with the Kuiper Airborne Observatory. The solid line represent the fit to the silicate feature by combining spectra from a variety of interplanetary dust particles. The dominant component is olivine with small amounts of pyroxene and layer lattice silicate (Bregman, J.D., Campins, H., Witteborn, F.C., Wooden, D.H., Rank, D.M., Allamandola, L.J., Cohen, M. and Tielens, A.G.G.M. 1987, Astron. Astrophys., **187**, 616).

Fig. 4.10. It shows the presence of several features superposed over the broad 10μm feature. The strong features are at 9.2, 10.0 and 11.2μm. The minor structures are at 10.5 and 11.9μm. The 11.2μm feature is due to crystalline olivine, as was seen earlier, in the Kuiper Airborne Observations on Comet Halley. Crystalline olivine has also a secondary peak at 10μm and a weak feature at 11.9μm. The broader 10μm maximum in the cometary spectra is characteristic of amorphous olivine. The characteristic signature of Pyroxene [(Mg,Fe)SiO$_3$] is the presence of a feature at 9.2μm. Therefore the cometary 9.2μm feature corresponds to amorphous Mg-rich Pyroxene. Crystalline pyroxene has also a feature near 9.3μm. Crystalline pyroxenes in general have a wide range of spectral features. The peaks at 10-11μm contribute to the width of total feature and the structure near 10.5μm. A profile fit to the 10 μm feature of several comets indicate the grain composition to be a mixture of amorphous and crystalline grains of olivine and pyroxene. It is rather difficult to determine the mass fraction

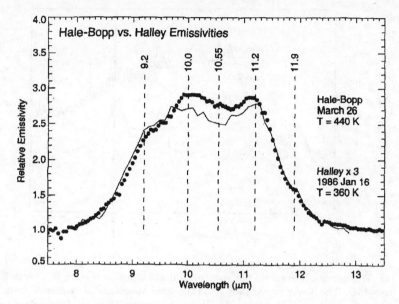

Fig. 4.10 Shows the shape of the emission feature around 10μm in Comets Halley (r=0.79AU, line) and Hale-Bopp (r=0.92AU, dots) with various spectral signatures (Hanner, M.S. and Bradley, J.P. 2004, In *Comets II*, Eds. M. Festou, H.U. Keller and W.A. Weaver, Univ Arizona Press,Tucson).

of crystalline to amorphous olivine. It appears to vary among the different comets and for the same comet at different heliocentric distances.

The spectra taken with Infrared Space Observatory on Comet Hale-Bopp at heliocentric distance, r=2.9AU showed the true nature of the grain material. The spectra show clearly five strongest emission features at 10, 19.5, 23.5, 27.5 and 33.5μm (Fig. 4.11). The wavelength of all the peaks correspond to Mg-rich crystalline olivine (forsterite, Mg_2SiO_4) when compared with the laboratory spectra (Fig. 4.11). Minor features present in the spectra are attributed to crystalline pyroxene [$(Mg,Fe)SiO_3$]. The infrared observations of comets have clearly shown that silicate grains in comets are Mg-rich. Models have been proposed to match the observed infrared spectra of Comet Hale-Bopp based on mixtures of silicate minerals. A good fit could be obtained with the following five emission components: blackbody 1(T=280K), blackbody 2(T=165K), Mg-rich olivine (forsterite, Cry Ol), Ortho pyroxene (Cry O-pyr) and amorphous (Am Pyr) silicates. The relative abundances of silicates are Cry Ol:Cry O-pyr:Am Pyr=0.22:0.08:0.70. The temperature of silicate grains used in the model is 210K.

ISO also revealed a striking similarity between the spectrum of Comet

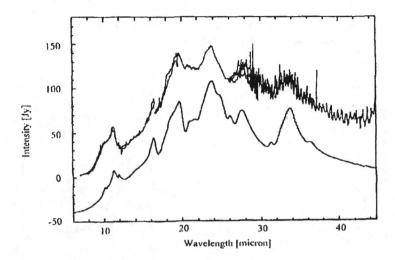

Fig. 4.11 The upper curve shows the observed silicate emission feature in Comet Hale-Bopp from ISO. The lower curve shows the modeled spectrum of forsterite from the laboratory data (Crovisier, J., Leech, K., Bockelee-Morvan, D. et al. 1997, *First workshop on Analytical Spectroscopy*, ESA SP-419, Eds. A.M. Heras, K. Leech, N.R. Trams and M. Perry. Nordwijk: ESTEC, p.137).

Hale-Bopp and that of dust disks around some young stars such as Herbig Ae/Be star HD100546 in the wavelength region 5 to 45μm. This can be seen from Fig. 4.12. A perfect match between the two indicate the common nature of the grain material.

4.9.2 Sizes of Grains

The impact detector on board Giotto spacecraft for Comet Halley measured the masses of the dust particles from which the sizes of the particles were deduced. The masses of individual dust grains varied from 10^{-16} to 10^{-11} gm. The deduced radii of cometary grains is around 0.2 to 2 μm. The presence of large number small-sized grains, ≤ 0.1 μm, which cannot be detected through observation made in the visible region was found to exist in Comet Halley in abundance. This result came out of the study of mass distribution of dust in the coma of Comet Halley (Fig. 4.13). The cumulative mass distribution function is consistent with a size distribution of the type $n(a) \propto a^{-\alpha}$ with $\alpha \sim 3.7$, which is similar to the values inferred from infrared observations and dynamics of dust tails of comets.

Larger size dust particles are also present in comets. This comes from

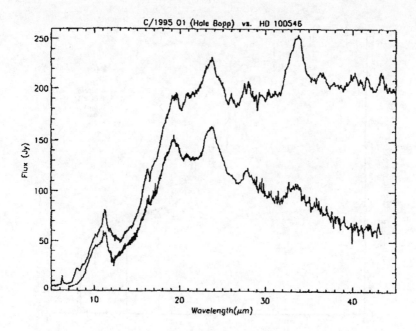

Fig. 4.12 Comparison between the ISO-SWS spectrum of HD100546 (top curve) with the corresponding spectrum from Comet Hale-Bopp (bottom curve). The striking resemblance between the two can clearly be seen (Malfait, K., Waelkens, C., Waters, L.B.F.M. et al. 1998, Astron. Astrophys., **332**, L25).

several considerations. Several comets do not possess a silicate feature. This could be attributed to difference in the particle sizes. The strength of the features around 10 and 20μm is a function of the particle size. The feature becomes weaker and disappear with an increase in size of the particles (Fig. 4.14). Therefore the lack of silicate feature could be attributed to the presence of larger size grains rather than to the deficiency of silicate material.

Comet Kohoutek showed the presence of anti-tail shortly after the perihelion passage. The infrared measurements carried out on Comet Kohoutek on the same day showed the presence of 10 μm feature in the coma and in the tail, but not in the anti-tail. The absence of 10 μm emission feature in the anti-tail immediately puts a lower limit to the particle size, a $\geq$ 5 μm. This conclusion is also consistent with the results based on dynamical considerations of the anti-tail which gives for the particle sizes, a $\geq$ 15μm.

The possibility that large particles can exist in the comae of comets came also from the laboratory experiments of clothrates-hydrate snows car-

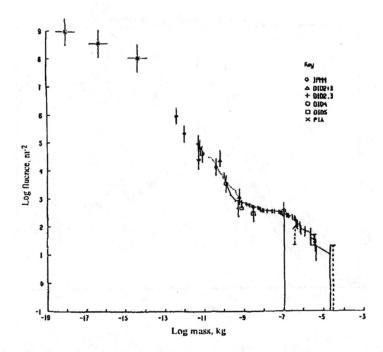

Fig. 4.13 A plot of dust particle fluence versus mass as measured by the DIDSY and PIA experiments on board the Giotto Halley probe (McDonnell, J.A.M., Lamy, P.L. and Pankiewicz et al. 1991, In *Comets in the Post-Halley Era*, Eds. R.L. Newburn, M. Neugebauer and J. Rahe, Kulwer Academic Publishers, p.1043: with kind permission of Springer Science and Business Media).

ried out under cometary conditions. These studies indicated the possible presence of icy-grain halo around the coma with particles of millimeter or centimeter sizes. Because of large sizes of the particles the thermal emission from such grains should be mostly in the radio spectral region. Infact the continuum emission at 3.71 cm was detected for the first time in Comet Kohoutek in 1974. Later, continuum emission at submillimeter wavelength has been detected from a large number of comets.

The impact detector on board Giotto spacecraft for Comet Halley also detected larger size dust particles (Fig. 4.13). The measured mass distribution of the dust particles for particle masses $> 10^{-9}$kg is found to be much flatter compared to smaller masses. This indicates that the total dust mass is mostly contributed by large size particles. Supportive evidence for the presence of large size dust particles comes from the study of meteor showers which are associated with comets. In addition dust trails seen in comets

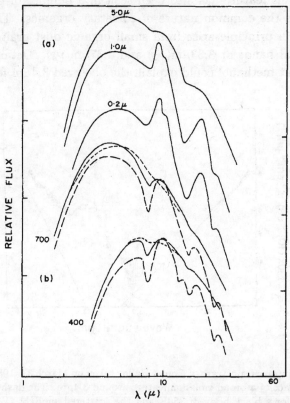

Fig. 4.14 Shape of the emission curve for different sizes, grain temperatures and for material of moon samples are shown. (a) for moon sample 12009, $T_g = 550K$ and sizes of 0.2, 1.0 and 5.0 μm. (b) for moon sample 14321, $T_g = 400K$ and 700K. Long-dashed, continuous and dashed curves refer to particle sizes of 2.5 and 10μ respectively (Krishna Swamy K.S. and Donn, B. 1979, Astron. J., 84, 692).

require millimeter to centimeter size dust particles. It is likely that the large size dust particles present in comets are the agglomerated interstellar dust grains in the solar nebula before they were incorporated into comets.

4.10 Organics

A broad emission feature was detected for the first time around 3.4μm in the spectra of Comet Halley, both from ground based and from spacecraft observations. This is shown in Fig. 4.15. Laboratory absorption spectra of several organic materials in gaseous and solid form generally show a strong feature around this wavelength and is attributed to C-H stretching vibrations. This feature has also been seen in a number of other comets. This indicates the common nature of cometary organics. The feature at 3.4μm could in principle arise from small organic dust grains. Methanol has vibrational bands at 3.33μm(ν_2) and 3.37μm(ν_9). Detailed modeling has shown that methanol could explain the observed 3.4μm feature.

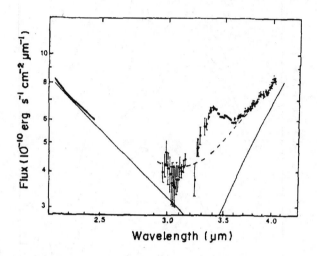

Fig. 4.15 The infrared spectrum of Comet Halley taken on March 31, 1986 (r = 1.17AU) shows the presence of a broad emission feature around 3.4μm. The dashed line refers to the continuum for a black body of 350K and the scattered sunlight (Wickramasinghe, D.T. and Allen, D.A. 1986, Nature, **323**, 44).

A weak feature at 3.29μm has been seen in several dusty comets. The observed spectra of Comet 109P/Swift-Tuttle at r=1.0AU is shown in Fig. 4.16. The feature at 3.29μm is a characteristic feature of aromatic bond. However, the other associated features at 6.2, 7.7, 8.6 and 11.3μm that are generally seen along with 3.29μm feature (AUIBs), is not present in the spectra of comets.

A new class of grains in Comet Halley called CHON particles came from the dust impact mass analyzer PUMA 1 and 2 on Vega and PIA on Giotto spacecrafts. Dust particles striking a silver target placed in front of the mass spectrometer generate a cloud of ions and the positive charge are mass analyzed which indicate its chemical composition. Several thousand mass spectra of dust particles were recorded by these instruments. These studies indicated broadly three classes of particles:(1) mostly made up of light elements such as H,C,N and O indicative of organic composition of

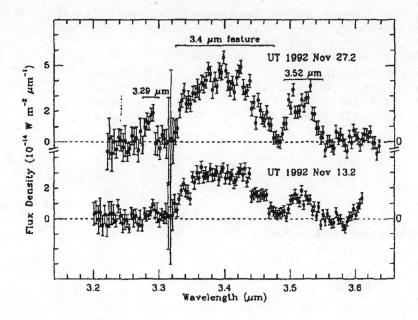

Fig. 4.16 The spectral feature of Comet 109P/Swift-Tuttle in the 3.4μm region (r=1.0AU) (Disanti, M.A., Mumma, M.J., Geballe, T.R. and Davis, J.K. 1995, Icarus, **116**, 1).

grain called CHON particles (2) Rocky-type (silicate with Mg, Si, Fe) grain and (3) a mixture of the above types of grains.

The results also indicated the presence of cometary dust particles composed of silicate core-with organic mantle structure. The spatial distribution of particles showed that silicate-Rock type of dust particles are relatively more abundant in the outer regions of the coma incontrast to CHON and mixed type of particles which are relatively more near to the nucleus. This indicates that organics are volatalized as they flow outwards from the nucleus releasing smaller size silicate-Rock dust grains.

The new class of grains, CHON particles, is made up of very complex organics. They are shown in Table 4.1. They provide evidence for the existence of many classes of aliphatic and aromatic hydrocarbons. They include unsaturated hydrocarbons such as pentyne, hexyne etc., nitrogen derivatives such as hydrogenic acid, aminoethylene etc., Heterocyclics with nitrogen such as purine and pyridine etc. HCN, H_2CO, allemine and pyrimidines and their derivatives which are biologically more significant appear to be present. Several peaks have been seen with positive ion cluster analyzer on Giotto spacecraft with an alternative mass difference of 14 and 16amu

Table 4.1 Types of organic molecules in Comet Halley Dust.

C – H – Compounds (Only high-molecular probable due to volatility; hints only to unsaturated)

Formula	Name
$HC \equiv C\,(CH_2)_2CH_3$	Pentyne
$HC \equiv C\,(CH_2)_3CH_3$	Hexyne
$H_2C = CH - CH = CH_2$	Butadiene
$H_2C = CH - CH_2 - CH = CH_2$	Pentadiene
(cyclopentene, cyclopentadiene structures)	Cyclopentene, cyclopentadiene
(cyclohexene, cyclohexadiene structures)	Cyclohexene, cyclohexadiene
(benzene, toluene structures)	Benzene, toluene

C – N – H – Compounds (Mostly of high extensity; also higher homologues possible)

Formula	Name
$H - C \equiv N$	Hydrocyanic acid
$H_3C - C \equiv N$	Ethanenitrile (acetonitrile)
$H_3C - CH_2 \equiv N$	Propanenitrile
$H_2C = N - H$	Iminomethane
$H_3C - CH = NH$	Iminoethane
$H_2C = CH - NH_2$	Aminoethene
$H_2C = CH - CH = NH$	Iminopropene
(pyrroline, pyrrole, imidazole structures)	Pyrroline, pyrrole, imidazole
(pyridine, pyrimidine structures)	Pyridine, pyrimidine (and derivatives)
(purine, adenine structures)	Purine, adenine

C – O – H – Compounds (Only very few hints to existence)

Formula	Name
$HC = OH$	Methanal (formaldehyde)
$H_3C - C = OH$	Ethanal (acetaldehyde)
$HCOOH \quad H_3C - COOH$	Methanoic (formic) and ethanoic 'acetic acid'

C – N – O – H – Compounds (Amino-, Imino-, Nitrile of ole, -ale, -keto- only probable with higher C-numbers of -anes, -enes, and -ines or cyclic aromates)

Formula	Name
$N \equiv C - OH \quad O = C = NH$	(Iso-) cyanic acid
$N \equiv C - CH_2 - OH$	Methanolnitrile
$HN = CH - CH = O$	Methanalimine

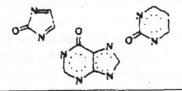

Oxyimidazole, oxypyrimidine

Xanthine

Kissel, J. and Krueger, F.R. 1987, Nature, **326**, 755.

as can be seen from Fig. 4.17. This could arise due to ion fragments CH_2 and O breaking off from the Formaldehyde polymer $(H_2CO)_n$, also known as Polyoxymethylene (POM). It is quite possible that CHON particles contain many more molecules than identified so far due to the fact that the impact velocity of the dust at the dust mass spectrometer was very high($\sim$ 78km/sec), which could possibly have destroyed many of them. The carbon content in cometary dust is close to 25% by weight. The gross charactor of the dust i.e silicate-Rocky type and CHON is similar to that of interstellar grains.

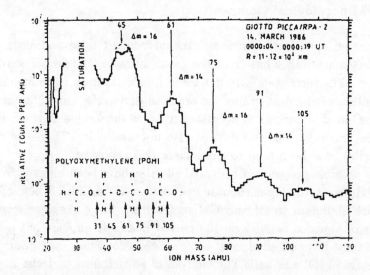

Fig. 4.17 The ion mass spectrum of Comet Halley observed on March 14, 1986 with the Giotto spacecraft at a distance of $11 - 12 \times 10^3$ from the nucleus. The peaks at regular intervals with alternate difference in mass number 14 and 16 have been interpreted as dissociation production of the molecule polyoxymethylene (Huebner, W.F., Boice, D.C., Sharp, C.M., Korph, A., Lin, R.P., Mitchell, D.C. and Reme, H. 1987, *Symposium on the Diversity and Similarities of Comets*, ESA-SP 278, p.163, See also Vanysek, V. and Wickramasinghe, N.C. 1975, Astrophys Space Sci., **33**, L19).

4.11 Water-Ice

Water-ice is the major component of the nucleas of a comet, the basis of Icy-conglomerate model of Whipple proposed in 1950's has been shown to be correct by all the subsequent observations. Hence among the dust particles in the coma, a significant fraction of water-ice particles could also be present. One way to detect such grains is to look for the absorption bands of water-ice in the infrared region namely features at 1.5, 2.2 and 3.1μm. The 3.1μm feature is much stronger than the other features. It is rather difficult to observe water-ice particles at smaller heliocentric distances where the grain temperature could be hot leading to shorter lifetime against sublimation. Therefore these features have been looked for in comets at larger heliocentric distances. The absorption features detected at 1.5 and 2.05 μm in Comet Hale-Bopp at r=7AU is attributed to that of water-ice grains.

4.12 Mineralogical Composition

As discussed already the mass spectra on board Giotto and Vega spacecrafts to Comet Halley has given the gross characteristic properties of dust particles. The next step is to get some information on the mineralogical composition of the dust grains. In order to derive the atomic abundances from the ion abundance it is necessary to know the ion yields which involve some uncertainty. The full data is also not available. Therefore various approaches have been used to circumvent this problem.

The possible presence of hydrated silicates have been inferred from the Comet Halley data. Mg-carbonate has also been identified. These minerals are quite abundant in CI and CM meteorites. Since they are formed as a result of aqueous activity on the parent bodies it implies that a similar process must have taken place at some stage in cometary dust. Carbonates are present in IDPs as well. The analysis of Fe-rich dust particles in Comet Halley has shown the presence of iron oxide, iron sulphide, Fe-rich silicates etc. There were also grains which had less Ni and other grains with more Ni, with a mean ratio of Ni/Fe $\approx$0.14. There was no indication of the presence of Ca-Al-rich grains and SiC grains.

From a knowledge of the abundances of various elements, it is possible to make a systematic analysis of its variation among the observed grains and make some broad classifications. For this purpose the cluster analysis method can be used. Cluster analysis is a statistical method of grouping a set of data points and to look for correlations among them. The method involves selection of several seed points with chosen distances. The data points are then assigned to these data points which form temporary clusters. Now these temporary clusters act as seed nuclei and the procedure is repeated till the changes are minimal. This method has been applied to classify Comet Halley's dust particles. In the study, the abundant inorganic elements Na, Mg, Al, Si, S, Ca and Fe have been considered. The result of these studies and with the observed distribution of Fe/(Fe+Mg), Mg-Fe-Si, Mg-Fe-O and Mg-Fe-S, it has been possible to characterize the particles into a few major mineral groups. The resulting major mineralogical composition of dust particles in Comet Halley is given Table 4.2. As can be seen from Table 4.2, the dominant component is the Mg-Si-O rich particles (Olivine,Pyroxene etc.). The other group comprising around 10% are the particles of Fe, Ni and sulphides. These particles could consist of Pyrrhotite ($Fe_{1-x}S$) and Pentlendite $(FeNi)_9S_8$ in analogy with IDPs. All these results indicate that the cometary dust is in an unequilibrated heterogeneous mixture of minerals containing both high and low temperature condensates. It is unlikely that all the components have a common origin.

Table 4.2 Estimated mineralogical composition of Comet Halley Dust.

Mineral group	Estimted proportion	Mineral chemistry	Possible minerals
Mg silicates	> 20%	Fe-poor, Ca-poor	Mg-rich pyroxene and/or olivine
Fe sulfides	~ 10%	some Ni-rich	pyrrhotite, pentlandite
Fe metal	1-2%	Ni-poor	kamacite
Fe oxide	< 1%		magnetite

Schulze, H., Kissel, J. and Jessberger, E.K. 1997, From *Star Dust to Planetesimals*, ASP Conference Series, Vol. 122, p.397, Eds. Y. Pendleton and A.G.G.M. Tielens, By the kind permission of the Astronomical Society of the Pacific Conference Series.

It is rather difficult to explain the presence of crystalline silicates in cometary dust. The amorphous-crystalline transition of silicate occurs at T~1000K. This is well above the temperature experienced by comets. It is possible that freshly formed crystalline material from the inner solar nebula is transported to regions where comets were formed. This implies extensive radial mixing of material in planetary disks. On this picture, comets formed much earlier in time will be rich in amorphous silicates as the original material had no time to be thermally processed. Therefore this could give rise to some variation in the crystalline to amorphous ratio between comets formed at different times. The other possibility is that they are presolar in origin. *This leads to the difficulty of explaining the nondetection of crystalline feature of silicates from interstellar space.* This could be explained if the temperature of pure Mg-silicate particles in interstellar space have temperature not high enough to show their emission feature. In general, crystalline silicates have lower grain temperature compared to amorphous silicate grains. This arises due to lack of FeO in crystalline silicate which reduces considerably the absorptivity of the grain. This results in lower grain temperature.

The emission from metals was seen in the Sun-grazing Comet Ikeya-Seki in 1965 which made it possible to deduce the elemental abundances. They were found to be similar to solar values. It is generally believed that when comets approach very close to the sun the dust grains which assume high temperature evaporates resulting in the release of metals. The *in-situ* measurements of Comet Halley made with impact ionization mass spectrometer on board Vega spacecraft have given a direct method of estimating the elemental composition of dust. The resulting abundances are very similar to the solar values except hydrogen (Fig. 4.18).

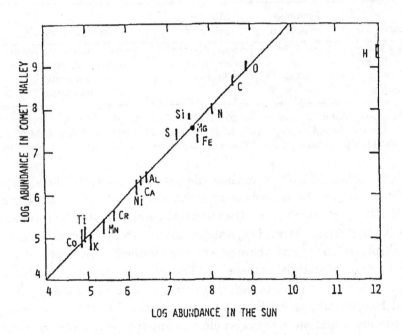

Fig. 4.18 A comparison of the elemental abundances in Comet Halley and in the Sun showing that they are very similar. The abundance refers to logN (H) = 12.0 (Delsemme, A.H. 1991, In *Comets in the Post-Halley Era*, Eds. R.L. Newburn, M. Neugebauer and J. Rahe, Kulwer Academic Publishers, p.377: with kind permission of Springer Science and Business Media).

4.13 Isotopic Studies

The study of the isotopic abundances of elements in cometary dust particles is important as it can provide clues to the physical conditions that prevailed at the time of formation of these objects. Fortunately most abundant elements namely H, C, N and O have many stable isotopes.

The best method of finding out the isotopic abundances of elements in the dust particles is to make direct measurement on the particles themselves, which was carried out by the various missions to Comet Halley. This was one of the main objectives of some of the instruments flown on the Giotto and Vega spacecrafts to Comet Halley in 1986. Several dust particles were found to be heavily depleted in ^{13}C and the ratio $^{12}C/^{13}C$ was found to go upto 5000. This is much larger than the normal value of 89. These particles are almost certainly presolar grains and indicative of circumstellar in origin. Therefore cometary dust particles may contain a fraction of circumstellar solids in addition to refractory interstellar material. The other measurement carried out with spacecrafts was the $^{18}O/^{16}O$ ratio which gave a value of 2.3×10^{-3} similar to the solar value. The results based on Giotto and Vega spacecrafts are very limited in extent. Therefore more direct isotopic measurements of cometary dust particles are needed for a better understanding of the origin of these dust particles. The comet dust sample return from NASA'S Stardust mission is an attempt in this direction. The spacecraft launched in February 1999 has already collected the dust particles from the Comet 81P/Wild 2 during its encounter with the comet in January 2004. After the return of the spacecraft in 2006, a detailed analysis of these particles in the laboratory should give direct information on the isotopic composition of elements in them and inturn on their origin.

Until more direct measurements become available it is necessary to fall back on the results derived from the study of coma gas as the dust particles could contribute a significant amount to the coma gas. This comes out of several considerations. As seen earlier the major component of cometary dust are the Rocky material, CHON particles, mixed type containing both the Rocky and CHON particles and icy particles. These particles coming out of the nucleus evaporate releasing the volatiles in the coma. The presence of CN jets in Comet Halley and CO as distributed specie in the coma of comets are also associated with the dust. Several new and long-period comets at r>4AU have a bright coma indicating the presence of substantial amount of water-ice grains. These icy grains could be released during the sublimation of the more volatile specie such as CO. However, for r≤3AU, the icy grains is unlikely to be observed in the coma. This is due to the fact that relatively high temperature of dirty ice particles(ice with impurities) for r≤3AU makes its lifetime quite small (a few hours or so) so that they evaporate within a few hundred kilometers from the nucleus. The most important isotopic ratio is the D/H ratio. The D/H ratio in cometary water is found to be $\approx 3 \times 10^{-4}$. This is a factor of around 10 larger than the local interstellar value $\sim 1.6 \times 10^{-5}$. However the cometary value is similar to the value $\sim 1.5 - 3 \times 10^{-4}$ found in 'hot core' of giant molecular clouds. Gaseous abundances in 'hot cores' of giant molecular clouds are found to be different from those in cold dense clouds. This difference could arise due to the evaporation of icy mantles from interstellar grains in 'hot cores'. Therefore the abundances in 'hot cores' may be directly comparable to cometary ices. The D/H ratio in cometary water is very similar to D/H ratio observed in meteorites and IDPs. The enrichment of deuterium in cometary material cannot be explained by processes in the solar nebula such as isotopic exchange reaction etc. It could have arisen through ion-molecule reactions or grain-surface processes prior to the solar nebula phase. The isotopic ratio of several other elments have been determined for several comets. So far most of the ratios measured, $^{12}C/^{13}C$, $^{16}O/^{18}O$ and $^{14}N/^{15}N$ are comparable to the terrestrial value.

In brief, the interpretation of dust observations in comets such as visible continuum, albedo, polarization and infrared continuum had given the gross property of the dust particles to be of silicate type of materials. The direct confirmation for the silicate nature of dust in comets came from infrared spectral studies. They have shown the mineralogy of cometary dust particles to be quite complex containing both amorphous and crystalline of the type olivine ($[(Mg,Fe)_2SiO_4]$) and pyroxene ($[(Mg,Fe)SiO_3]$). The detailed structure seen in the ISO spectra of Comet Hale-Bopp in the wavelength region 5 to 45μm is found to correspond to each of the features present in the laboratory spectra of forsterite (Mg_2SiO_4). The ISO spectra of Comet Hale-Bopp is also found to be very similar to the ISO spectra of circumstellar dust around evolved stars indicating the similar nature of the silicate material in both of them. The detection of a broad emission feature at 3.4μm in Comet Halley for the first time and later on in other comets is attributed to the C-H stretching mode. This showed the common nature of organics present in comets. These results are consistent with the *in-situ* measurements of the composition of the dust carried out with Giotto and Vega spacecrafts on Comet Halley. These measurements showed three types of dust particles, namely dust with Rock-forming elements (Mg, Si, Ca, Fe), CHON (organic containing H, C, N and O) and a mixture of these two types of particles. Icy grains are also present. The *in-situ* measurements also showed the presence of very small and very large sized grains in Comet Halley. The isotopic ratio $^{12}C/^{13}C$ of several dust particles in Comet Halley extends upto 5000, indicating that they are presolar in origin. The D/H ratio in cometary water and in 'hot cores' of giant molecular clouds are very similar. Therefore icy mantles of interstellar grains in 'hot cores' may be directly comparable to cometary ices. The laboratory analysis of dust samples of Comet 81P/Wild 2 collected by NASA'S Stardust mission should give a better understanding of the mineralogy of cometary dust and inturn its origin. The mineralogy of cometary dust is quite similar to chondritic aggregate interplanetary dust particles which are believed to come from comets.

Lastly, the study of comets is vital for an understanding of some of the basic and fundamental issues. The pristine nature of its material should help in the understanding of the origin of the solar system. The organic material deposited during the early stages on the Earth could have possibly led to life on the Earth. It could have also deposited water on the Earth. The laboratory analysis of the Stardust mission data from Comet 81P/Wild 2 should provide a direct link to dust in the interstellar medium.

References

The following references cover most of the essentials

Festou, M., Keller, H.U. and Weaver, H.A. (Eds.) 2004, *Comets II*, University of Arizona Press.

Grun. E. and Jessberger, E. 1990, In *Physics and Chemistry of Comets*, Ed. W.F. Huebner, Springer-Verlag, p.113.

Krishna Swamy, K.S. 1997, *Physics of Comets*, World Scientific Publishing Co., Singapore.

Newburn Jr, R.L., Neugebauer, M. and Rahe, J. (Eds.) 1991, *Comets in the Post-Halley Era*, Vols.1 and 2, Kluwer Academic Publishers.

Sekanina, Z., Hanner, M.S., Jessberger, E.K. and Fomenkova, M.N. 2001, In *Interplanetary Dust*, Eds. E. Grun, Bo.A.S. Gustafson, S. Dermott and H. Fechtig, Springer-Verlag, p.95.

The theory of dynamics of dust tails is worked out in detail in these papers
Finson, M.L. and Probstein, R.F. 1968, Astrophys. J., **154**, p.327, p.353.

Later Papers
Combi, M.R., Kabin, K., DeZeeuw, D.L., Gombosi, T.I. and Powell, K.G. 1999, Earth Moon and Planets, **79**, 275.
Crifo, J.F. 1991. In *Comets in Post-Halley Era*, Eds. R.L. Newburn Jr., M. Neugebauer and J. Rahe, Kluwer Academic Publishers, Vol.2, p.937.
Fulle, M. 1989 Astron. Astrophys., **217**, 283.

Dust Trail
Eaton, N, Davis, J.K. and Green, S.F. 1984, Mon. Not. Roy. Astron. Soc., **211**, 15.
Sykes, M.V., Lebofsky. L.A., Hunten, D.M. and Low, F.J. 1986, Science, **232**, 1115.

Detection in the visible
Ishiguro, M., Watanabe, J., Usui, F., Tanigawa, T., Kinoshita, D., Suzuki, J., Nakamura, R., Ueno, M. and Mukai, T. 2002, Astrophys. J., **572**, L117.

Visible Continuum.
Early Papers
Liller, W. 1960, Astrophys. J., **132**, 867.
Remby-Battiau, L. 1964, Acad. r. Belg. Bull. el. Sci. 5eme Ser. **50**, 74.

Later Paper
Jewitt, D.C. and Meech, K.J. 1986, Astrophys. J., **310**, 937.

Polarization.
Early Work
Bappu, M.K.V. and Sinvhal, S.D. 1960, Mon. Not. Roy. Astron. Soc., **120**, 152.

Later Work
Dollfus, A. 1989, Astron. Astrophys., **213**, 469.
Kelley, M.S., Woodward, C.E., Jones, T.J., Reach, W.T. and Johnson, J. 2004., Astron. J., **127**, 2398.
Levasseur-Regourd, A.C., Hadamcik, E. and Renard, J.B. 1996, Astron. Astrophys., **313**, 327.
Mukai, T., Mukai, S. and Kikuchi, S. 1987, Astron. Astrophys., **187**, 650.

IR Observations.
The First IR measurement made on Comet Ikeya-Seki is reported by
Becklin, E.E and Westphal, J.A. 1966, Astrophys. J., **145**, 445.

First detection of $10\mu m$ emission feature in Comet Bennett
Maas, R, Ney, E.P and Woolf, N.J. 1970, Astrophys. J., **160**, L101.

ISO Observation
Crovisier, J., Brooke, T.Y., Leech, K., Bockelee-Morvan, D., Lellouch, E., Hanner, M.S., Altieri, B., Keller, H.U., Lim, T., Encrenaz, T., Salama, F., Griffen, M., de Graauw, T., van Dishoeck, E. and Knacke, R.F. 2000, In *Thermal Emission Spectroscopy and Analysis of Dust Disks and Regoliths*, Eds. M.L. Sitko, A.L. Sprague and D.K. Lynch, ASP Conference Ser. Vol.196, p.109.

Other Papers
Donn, B., Krishna Swamy, K.S. and Hunter, C. 1970, Astrophys. J., **160**, 353.
Hanner, M.S., Gehrz, R.D., Harker, D.E., Hayward, T.L., Lynch, D.E., Mason, C.G., Russel, R.W., Williams, D.M., Wooden, D.H. and Woodward, C.E. 1999, Earth Moon and Planets, **79**, 247.
Krishna Swamy, K.S. and Donn, B. 1968, Astrophys. J., **153**, 291.
Krishna Swamy, K.S., Sandford, S.A., Allamandola, L.J., Witteborn, F.C. and Bregman, J.D. 1989, Astrophys. J., **340**, 537.
Wooden, D.H. 2002, Earth Moon and Planets, **89**, 247.
Wooden, D.H., Harker, D.E., Woodward, C.E., Koike, C., Witteborn, F.C., McCarthy, M.C. and Butner, H.M. 1999, Astrophys. J., **517**, 1034.

Albedo
Hanner, M.S. and Newburn, R.L. 1989, Astron. J., **97**, 254.
O'Dell. C.R. 1971, Astrophys. J., **166**, 675.

Sizes of Grains.
in-situ
Mazets, E.P., Aptekar, R.L., Golenetskii, S.V., Guryan, Y.A., Dyachkov, A.V., Ilynskii, V.N., Panov, V.N., Petrov, G.G., Savvia, A.V., Sagdeev, R.Z., Sokolov, I.A., Khavenson, N.G., Shapiro, V.D. and Shevchenko, V.I. 1986, Nature, **321**, 276.
McDonnell, J.A.M., Lamy, P.L. and Pankiewicz, G.S. 1991, In *Comets in the Post-Halley Era*, Eds. R.L. Newburn, M. Neugebauer and J. Rahe.Kluwer Academic Publishers, p.1043.

Radio Observations
Hobbs, R.W., Maran, S.P., Brandt, J.C., Webster, W.J. and Krishna Swamy, K.S. 1975, Astrophys. J., **201**, 749.

Organics
Disanti, M.A., Mumma, M.J., Geballe, T.R. and Davis, J.K. 1995, Icarus, **116**, 1.
Mumma, M.J. 1997, In *From Stardust to Planetesimals*, Eds. Y.J. Pendleton and A.G.G.M. Tielens, ASP Conference Ser. Vol.122, p.369.

Mineralogy
Jessberger, E.K. 1999, Space. Sci. Rev., **90**, 91.

Schulze, H., Kissel, J. and Jessberger, E.K. 1997, In *Stardust to Planetesimals*, Eds. Y.J. Pendleton and A.G.G.M. Tielens. ASP Conference Ser. Vol.122, p.397.

Isotopic Studies
Altwegg, K. and Bockelee-Moravan, D. 2003, Space. Sci. Rev., **106**, 139.
Eberhardt, P., Reber, M., Krankowsky, D. and Hodges, R.R. 1995, Astron. Astrophys., **302**, 301.

Paper on HD100546
Bouwman, J., de Koter, A., Dominik, C. and Waters, L.B.F.M. 2003, Astron. Astrophys., **401**, 577.

Chapter 5

Interplanetary Dust

5.1 Introduction

An important component of dust in the universe is the Interplanetary Dust Particles which are collected at stratosphere altitude (∼20km on the Earth). In addition some of the matter which survive the passage of the atmosphere and fall on the surface of the Earth called Meteorites are also an important source of information on the dust component of the universe. In this chapter these aspects will be discussed.

5.2 Interplanetary Dust Particles

5.2.1 *Morphology, Structure and Chemical Composition*

Various methods have been used for the collection of particles from the Earth's upper atmosphere using recoverable rockets, balloons and aircrafts. These particles are generally called Interplanetary Dust Particles (IDPs). The most extensive collection of IDPs is carried out with NASA U2 aircraft. The particles collected are subjected to a thorough laboratory investigation for the study of their morphology, structure and chemical properties.

Typical IDPs are aggregates of large number of smaller particles. This can be seen from Fig. 5.1. The diameter of aggregates varies from 5 to 25μm. Typical grain sizes within the aggregates are 0.1-0.5μm. IDPs have various shapes and highly fluffy. They are structurally weak. Enstatite (pyroxene, $MgSiO_3$) IDPs exhibit several interesting morphologies and microstructures such as rods, ribbons and platelets.

The most common IDPs is called chondritic because of the observed relative abundances of the major rock-forming elements in these IDPs are

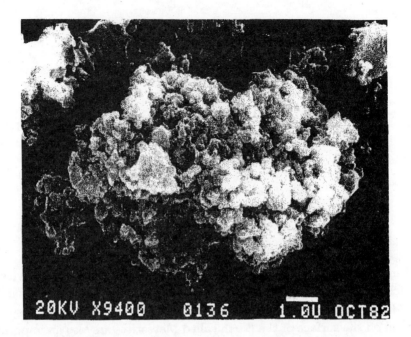

Fig. 5.1 Scanning electron micrograph of a typical chondritic interplanetary dust particle. It shows clearly the microstructure and high porosity of the particle. The grains of submicron sizes are mostly GEMS and carbonaceous material. The grains of micron sizes contain single crystals of forsterite, enstatite and iron-nickel sulphides. Bar is $1\mu m$ in length (Hanner, M.S. and Bradley, J.P. 2004, In *Comets II*, Eds. M. Festou, H.U. Keller and W.A. Weaver, University Arizona Press, Tucson).

similar to those of primitive meteorites called carbonaceous chondrites (Sec. 5.3) The structure of IDPs are of two kinds. They are chondritic smooth and chondritic porous types. The measured density for the particles of chondritic composition lies between 0.7 and 2.2 gm/cm^3. This leads to a poracity ∼40%. The wide variation in the observed density of IDPs is due to their poracity rather than to differences in elemental composition. The albedo in the visible region derived from spectral reflection measurements lie in the range 5-15%. The low value for the albedo arise due to the porous structure and to the absorbing material.

Early studies on chondritic IDPs had indicated the presence of various kinds of minerals such as Pyroxene, Olivine, Layer-Lattice silicate, Carbonaceous material, Iron-rich Sulphides etc. However from infrared measurement of IDPs it is possible to broadly classify them into subgroups. The infrared spectra between 2.5 and 25μm of a large number of IDPs shows the presence of a strong 10μm silicate absorption feature and possibly the

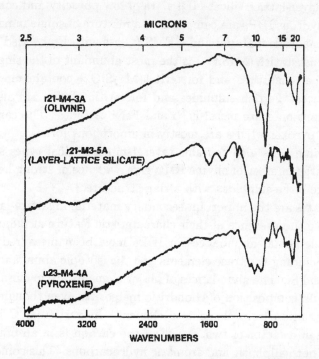

Fig. 5.2 Representative infrared spectra from three main IDP infrared classes. From top to bottom the spectra are dominated by characteristic features of olivines, layer-lattice silicates and pyroxenes, respectively. In addition to 10 and 20μm silicate features, features at 3.0 and 6.0μm due to hydration and a feature near 6.9μm due to carbonates are also present (Sandford, S.A. and Walker, R.M. 1985, Astrophys. J., **291**, 838: reproduced by permission of the AAS).

3.4μm feature, as can be seen from Fig. 5.2. Based on the structure of the 10μm silicate feature three distinct mineralogical classes of chondritic IDPs are identified. These three are referred to as the olivine, pyroxene and layer-lattice silicate class. As the name indicates the spectra of olivine and pyroxene class in IDPs match with those of terrestrial olivine and pyroxene. The spectra of layer-lattice silicate group is similar to spectra of minerals containing layer-lattice structure. In each group one particular type of mineral dominate. For example in olivine group, olivine is the dominating component. In addition they contain other minerals with less abundances. The layer-lattice group is different from the other two groups in their characteristic internal structure. The olivine and pyroxene groups are also known together as anhydrous IDPs and layer-lattice silicates as hydrous IDPs. In general, anhydrous IDPs consists of higher porous aggregates. But layer-lattice silicate IDPs are of low poracity and compact.

The anhydrous IDPs are composed of a mixture of single mineral grains, glass, carbonaceous material and GEMS (Glass with Embedded Metal and Sulphides). Enstatite ($MgSiO_3$) is the most abundant of the single mineral grains. Other enstatites and forsterites(Mg_2SiO_4) contain more Mn and Cr abundances. Fe-rich sulphides and forsteritic olivine are also present. Less abundant ones are metal-FeNi and FeNi carbides. The carbonaceous material in pyroxene IDPs are mostly in amorphous form.

The olivine class also contains several mineralogical types seen in pyroxenes. Large number of olivine IDPs give evidence of strong heating with the presence of Fe-sulphides with a ring structure.

Carbonates are the important secondary material in lattice-layer silicate IDPs, due to the presence of their characteristic feature at 6.8μm.

Major elemental abundances in IDPs have been measured with high precision including the trace elements and the isotopic abundances of H, C, N, O, Mg and Si. The abundances of major elements are in close agreement with the bulk composition of chondritic meteorites even though there may be some variation from element to element. In particular C is found to be more than a factor of two. Much of the carbon is in an organic phase containing both aliphatic and aromatic hydrocarbons. This comes from the studies based on X-ray absorption spectroscopy as well as the presence of C-H stretch absorption feature at 3.4μm and the feature at 5.85μm attributed to C=O groups. The C=H function groups are present which is consistent with the nature of the CHON particles detected in Comet Halley.

IDPs have been studied with Laser Raman microbe technique. The results show the presence of Raman bands and a broad emission in the red region of the spectrum attributed to luminescence. The Raman spectra do not show the characteristic feature of silicate but instead the spectra show strong double peak feature at $\sim 1360 cm^{-1}$ and $1600 cm^{-1}$, corresponding to 7.38 and 6.25μm. This is shown in Fig. 5.3. These are due to C-C vibrations of carbon. A comparison of these features with laboratory studied synthesized materials indicate that the carbonaceous material contains crystalline units smaller than 25Å. This indicates the presence of PAHs of some kind.

Many of the IDPs are found to contain large deuterium enrichments (upto D/H$\sim 8 \times 10^{-3}$), about a factor of around 800-1000 larger compared to mean local interstellar medium D/H values. The deuterium enrichment seems to correlate with carbon abundance suggesting a carbonaceous carrier. Only IDPs of the hydrated silicate variety and the pyroxene were found

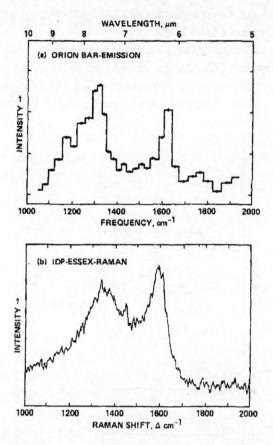

Fig. 5.3 Comparison between the 6.2 and 7.4μm interstellar emission features and the Raman spectrum of an IDP. The similarity in the spectra suggests aromatic compounds are responsible for both set of features (Sandford, S.A. 1989, In *Interstellar Dust*, IAU Symposium No. 135, Eds. L.J. Allamandola and A.G.G.M. Tielens, Kulwer Academic Publishers, p.403 (figure taken by J.P. Bradley): with kind permission of Springer Science and Business Media).

to exhibit deuterium isotopic anamalies but not olivine type IDPs. Substantial enrichments of ^{15}N and ^{16}O have also been seen in IDPs. Therefore the major isotopic anamolies found in IDPs are those of H, N and O.

The major form of noncrystalline silicate in IDPs are the GEMS. GEMS are submicron-sized spheroidal grains containing nanometer-sized FeNi metal and Fe-rich sulphide grains embedded in silicate glass. The bulk composition of GEMS are approximately chondritic for the major rock forming elements. The GEMS show evidence for the exposure to larger doses of ionizing radiation indicated by the presence of eroded surfaces, oxygen enrichments etc. This indicates that the exposure must have occurred prior to the accretion of the IDPs.

The infrared spectra obtained from thin sections of two pyrrhotite-rich IDPs, L2011*B6 and U2012A-2J, are shown in Fig. 5.4. L2011*B6 is a fragment from the deuterium rich cluster IDP. They show a strong and broad feature around 23μm, which is identified with FeS (troilite).

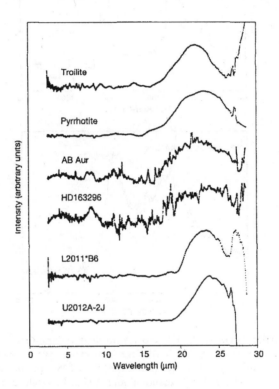

Fig. 5.4 Infrared spectra of IDPs (U2012A-2J and L2011*B6), Herbig Ae/Be stars (AB Aur and HD 163296) and FeS (troilite). The residual spectra for Ab Aur and HD 163296 is shown after subtracting the best model fit to the observed spectra. The dust components used in the model are, glassy silicate, forsterite, carbonaceous material, metallic iron and water ice. The mineral of U2012A-2J is dominated by pyrrhotite and that of L2011*B6 by low-NiFi-sulphides (Keller, L.P., Hony, S., Bradley, J.P. et al. 2002, Nature, **417**, 148).

The above discussion pertained to the particles collected from the stratosphere. Another class of IDPs generally called as micrometeorites refer to the IDPs collected from Antarctic and Greenland ice. Large number of particles of this type have also been studied. Olivine and pyroxene grains are the most common anhydrous silicates in micrometeorites. Magnetite and hydrous Mg-Fe silicates such as serpentine [$(Mg,Fe)_6Si_4O_{10}(OH)_8$] and saponite are also present. Minor components are Ca-rich pyroxenes, feldspars, Fe-Ni sulphides and metal, Mg-Fe hydroxides etc. Minerology, mineral chemistry and the presence of refractory minerals are similar to those of CM-type and CR-type carbonaceous chondrites. But there are considerable differences as well. There is considerable similarity between micrometeorites and IDPs with regard to their mineralogical and chemical composition. However GEMS seen in IDPs does not appear to be present in micrometeorites.

5.2.2 Origin

The direct proof for the extraterrestrial nature of Interplanetary Dust Particles comes from several considerations. IDPs have been exposed to solar wind and cosmic rays for sufficiently long time to accumulate large amounts of noble gases and spallogenic isotopes. For example, He and Ne contents are in excess compared to the normal values and also the isotopic abundances of these elements are similar to those of solar energetic particles. The presence of ^{10}Be indicates the production through cosmic ray bombardment. There is enrichment of deuterium as well (i.e D/H ratio).

The major source of IDPs are asteroids and comets. There are several ways to differentiate between the two sources. One method is from the study of Solar-flare tracks present in IDPs.

When dust particles from comets and asteroids of sizes larger than $1\mu m$ are injected into the solar system they are acted upon by the gravitational field of the Sun and the Poynting-Robertson drag force. Due to Poynting-Robertson effect, the particles spiral in orbits and become progressively more circular. As they are spirling in they accumulate solar flare tracks in their mineral components. The track density within IDPs depends upon several factors such as space exposure time, its distance from the sun and the flux of track-producing solar-flare nuclei. Computer simulations have been carried out for the expected track density distribution from grains produced by comets and asteroids. They show a very narrow distribution at higher track densities for grains coming from asteroids while grains from comets show a flatter distribution. This is because comets are in highly elliptical orbits that may cross the Earth orbit. Therefore cometary particles with short exposure time and low track densities can be collected. However particles from asteroids are nearly in circular orbits and hence are exposed for a longer time interval before their orbits evolve and finally cross the

Earths orbit. This results in narrow range of track densities. Particles with track densities above a certain value have probably been exposed to solar flare tracks prior to their entry into interplanetary medium. They are likely to be of asteroidal origin. Particles with track densities less than the above value is more likely of cometary source rather than from Earth-crossing asteroids. The observed track densities are found to lie within the ranges of the calculated values.

It is also possible in principle, from the above dynamical study to distinguish between the two sources from their velocity of atmospheric entry. These results show that in general cometary particles have higher entry velocities compared to those of asteroidal particles with a small velocity overlap between the two groups of particles. Therefore cometary particles are expected to be heated more compared to asteroidal particles.

The majority of the particles collected of the type layer-lattice silicate and pyroxene have not been heated to high temperature during atmospheric entry. This indicate that these particles arrive at the top of the atmosphere with low geocentric encounter velocities. This implies that the trajectories of these particles must be relatively circular or prograde orbits. Therefore these particles cannot come from Earth-crossing comets or asteroids. More likely the source of pyroxene and layer-lattice silicate particles are from comets and asteroids with low orbit inclinations. On the other hand, the collected olivine-rich IDPs appears to have been heated to high temperature during atmospheric entry. Therefore these particles have to be captured from a more eccentric orbits. Therefore olivine-rich IDPs are of cometary origin. The chemical and mineralogical information indicate that pyroxene-rich IDPs and olivine IDPs are from comets (Fig. 5.5) and layer-lattice IDPs are from asteroids.

Supporting evidence for the cometary origin comes from the presence of meteor showers as well as the factual information that the vaporized material of the comet at each apparition is dispersed into the interplanetary medium some of which may find its way on to the Earth.

Many IDPs appear to be more primitive in composition compared to meteorites based on mineralogical, chemical and texture properties. They also have substantially higher carbon and volatile element abundances than in meteorite. All the observations seem to suggest that negligible parent body alteration has taken place in IDPs compared to those of meteorites.

As mentioned earlier, many of the IDPs are found to contain large deuterium enrichments. The origin of deuterium excess is somewhat uncertain. Deuterium enrichment via chemical reactions require very low tempera-

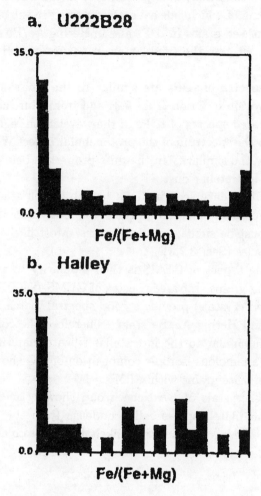

Fig. 5.5 Histograms comparing the submicrometer-scale distribution of Fe/(Fe+Mg) in IDPs with Comet Halley dust. The horizontal scale ranges from 0 to 1. (a) Chondritic porous pyroxene IDP U222B28 which contains abundant Mg-rich silicates, mostly enstatite ($MgSiO_3$) and some forsterite ($MgSiO_4$) result in a large number of Fe/(Fe+Mg) data points near zero. (b) Halley (PUMA) mass spectrometry data (Bradley, J.P., Snow, T.P., Brownlee, D.E. and Hanner, M.S. 1999, In *Solid Interstellar Matter: The ISO Revolution*, Eds. L. d'Hendecourt, C. Joblin and A. Jones, Springer-Verlag, p.297: with kind permission of Springer Science and Business Media).

tures, $T \leq 100K$. However at such low temperatures the reaction rate is slow. But the reaction rate for ion-molecule reactions are high. Therefore the source of enrichment of deuterium in IDPs is suspected to be due to ion-molecule reactions taking place in dense, cold interstellar clouds. Molecules formed under these conditions were presumably incorporated into interstellar dust grains some of which subsequently survived in the solar nebula.

The presence of FeS grains in IDPs and in Herbig Ae/Be stars (AB Aur and HD 163296), suggest the presolar grain nature of FeS seen in IDPs (Fig. 5.4).

The Raman spectra of IDPs are similar to the emission spectrum of the Bar nebular region in Orion (Fig. 5.3) and from Murchison meteorite. Similarly, the infrared spectra of IDPs in the wavelength region 1 to 14μm, matches well with the spectrum of the protonebular object W33A. All these observations show the similarity in the dust properties between IDPs, meteoritic dust and interstellar dust.

The photoluminescence seen in IDPs has also been seen from several other dusty environments like, HII regions, planetary nebulae etc. Therefore the source must be similar in all the cases. Most likely they are from silicon nano-particles (Sec. 3.7.3).

The interesting aspect of GEMS is their similarity to amorphous silicates of interstellar grains. Infrared spectra of GEMS-rich IDPs (containing GEMS and silicate crystals) provide a good spectral match with cometary silicate spectra and Herbig Ae/Be stars. Therefore the common components of IDPs are similar to the interstellar silicate material (Table 5.1). The detection of anamolous isotopic composition of ^{17}O show the presence of circumstellar silicate grains within IDPs.

The Enstatite crystals of pyroxene group showing morphologies like rods, ribbons and platelets, generally condense from the vapour phase. Therefore they would have formed most likely from a presolar environment.

5.3 Meteorites

The meteorites are bodies which survive the passage through the Earth's atmosphere and fall on the ground. Only the largest objects create large impact craters on hitting the ground. Large number of meteorites are found in Antartica as they are preserved over long periods of time. Also they are scattered over large distances due to fragmentation of the body passing through the Earth's atmosphere.

Table 5.1 Similarity between GEMS and interstellar silicates.

Properties of GEMS	Properties of Interstellar Silicates
0.1-0.5 μm diameter	0.1-0.5 μm diameter
"Amorphous" silicate (glass)	Amorphous silicates (Mathis, 1993)
Contain Nanocrystals of FeNi metal	Hypothesized to contain nanocrystals of metal ($<$ 10% vol.) which explains interstellar grain alignment and optical properties (i.e. "dirty silicates") (Jones and Spitzer, 1967)
Approximately chondritic (solar) compositions (except for C)	Approximately solar compositions except for C (Mathis 1993)
Systematically depleted in S	Systematicaly depleted in S (Martin, 1995)
Heavily irradiated	Inevitably irradiated
Ubiquitous in cometary IDP's	Expected to be perserved in cometary IDPs (Greenberg, 1982)

Bradley et al. 1997, In *From Stardust to Planetesimals*, Eds. Y.J. Pedlenton and A.G.G.M. Tielens, ASP Conference Series, Vol. 122, p.217: By the kind permission of the Astronomical Society of the Pacific Conference Series.

5.3.1 *Morphology, Structure and Chemical Composition*

Large number of meteorites have been studied over the years with regard to their structure, morphology, chemical composition, origin etc. Of particular interest are the two meteorites, Murchison and Allende. Murchison meteorite is a CM chondrite that fell near Murchison in Australia. Allende meteorite is a CV chondrite that fell near Pueblito de Allende in Mexico. In general the material available from a meteorite for laboratory studies is very limited in extent because of their small size. But Murchison and Allande meteorites were very large in size and therefore large amount of material was available. This gave an opportunity to do extensive investigation of various kinds in the laboratory on these two meteorites which otherwise is not possible. In addition Allende meteorite had the advantage of finding it within a short time of its fall. This gave the chance to study the material in its true form with minimum loss of some of the volatiles. Also Allende meteorite had abundant cm-size inclusions (see below).

Meteorites are broadly classified into three groups based on their chemical and mineralogical components. They are stony, iron and stony-iron metorites (Table 5.2). The iron-type meteorites are essentially pieces of

Table 5.2 Main Meteoritic Groups.

Group	Main Characterization	Main Components
Iron meteorites	More than 90% metal	Nickel-Iron
Stony meteorites:		
(i) Chondrites	More than 75% story material chondrules	Silicates, solar in composition except volatile elements
(ii) Achondrites	No chondrules	Silicates, differ from solar composition
Stony-Iron meteorites:		
(i) Pallasites	Olivine in Fe-Ni alloy network	Olivine, metal
(ii) Mesosiderites	silicate and metal grown together	silicate, metal

metal containing Fe-Ni alloy. The stony-iron type contain mixtures of stony material and metal. The stony-type consists mostly silicate type material. The chondrites belong to the class of the stony-type meteorites which are found to contain mm-size spheroidal globules called Chondrules. Chondrules are essentially silicate mineral assemblages typically 0.1 to 1.0mm in size which had an independent existence prior to the formation of the meteorite. They have a wide variety of internal structures but generally consist of silicate grains enclosed in glass or in crystals. The chondrites have a composition which is remarkably similar to that of the photosphere of the sun, for all but the most volatile elements. Achondrites differ considerably from the composition of the sun.

The three broad classes of meteorites are further sub-classified according to the crystal structure, mineral content and the state of metamorphism. The subgroup in chondrites, namely CI, CM, CO and CV chondrites, are characterized by relatively high carbon content and are collectively termed as 'Carbonaceous Chondrites'. Carbonaceous chondrites are well known for having mm-sized refractory inclusions (Ca- and Al-rich inclusions,CAI). The mineralogy is dominated primarily by oxides and silicates. CAI range in shape from highly irregular to nearly spheroidal and in colour from white to pink. They are most abundant in carbonaceous chondrites.

Large number of minerals of various types have been identified in meteorites. Many of them are present in trace amount. The important minerals are olivine and pyroxene formed out of Fe, Mg and Ca-silicates. There are also some pyroxene minerals with Na, Al etc. The sulphide in chondrites is almost always FeS (troilite). These appear to exist as small, widely dispersed grains surrounded by silicates. In addition a variety of around 400 types of organics have been identified in Murchison meteorite. They include substances like aliphatic and aromatic hydrocarbons, amino acids, amines and amides, carbonic acid and so on. Several of these are common with biological systems.

The presolar spherical graphite particles have two well defined structural types: namely Cauliflower-like type and Onion-like type (Figs. 5.6, 5.7). In the Onion-like type, there is a core of nanocrystalline carbon made up of aromatic carbon units likely to have condensed from a vapour phase. This core is then surrounded by layers of well grown graphitized carbon. In the Cauliflower-like type of graphite grains, there is no long range continuity but arranged in small chunks indicating the growth pattern with several surface nucleation centers. Some of the experiments carried out in the laboratory has shown the growth pattern of cauliflower-like structure for the particles.

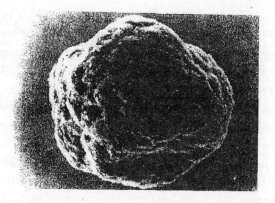

Fig. 5.6 Scanning electron microscope micrograph of a Cauliflower-type graphite spherule from the Murchison meteorite. It is a presolar grain of $10\mu m$ in size and with $^{12}C/^{13}C$ of 43.5 (Bernatowicz, T.J. 1997, In *From Stardust to Planetesimals*, Eds. Y.J. Pendleton and A.G.G.M. Tielens, ASP Conference Ser., Vol.122, p.227: By the kind permission of the Astronomical Society of the Pacific Conference Series).

5.3.2 *Presolar Grains*

The extraction and the study of presolar grains in meteorites has given rich dividends. This comes out of the study of the anomalous isotopic composi-

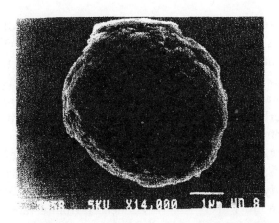

Fig. 5.7 Scanning electron microscope micrograph of a Onion-type graphite spherule from the Murchison meteorite. It shows clearly the surface shells of graphite. It is a presolar grain of 5μm in size and with $^{12}C/^{13}C$ of 238 (Bernatowicz, T.J. 1997, In *From Stardust to Planetesimals*, Eds. Y.J. Pendleton and A.G.G.M. Tielens, ASP Conf.Ser. Vol.122, p.227: By the kind permission of the Astronomical Society of the Pacific Conference Series).

tion of several noble gases such as Ne, Xe etc. seen mostly in the inclusions. Here the isotopic anomalies refer to the values in comparison to that of average solar system isotopic composition for a given element. The noble gases are so rare any deviation from the average value can be quite significant. It is hard to explain these observed isotopic anomalies in terms of the local production mechanisms such as irradiation, mass fractionation process or known radioactive decay. Hence these anomalies represent materials injected into the solar system from nucleosynthetic sites within a relatively short time of their synthesis. It is highly unlikely that all the anomalies seen in meteorites have a common origin and therefore must have different nucleosynthetic origin. Therefore laboratory study of these presolar grains can provide important information on the sites of nucleosynthesis, the possible origin of condensation of these grains etc.

The presolar grains identified in meteorites are mostly carbon- bearing, namely diamond, graphite and silicon carbide.

Presolar diamond detected in meteorites came from the study of enrichment of isotopic composition of Xe; i.e. from the study of isotopes ^{134}Xe, ^{136}Xe and ^{124}Xe, ^{126}Xe. Isotopic anomalies have also been seen in the elements Ba, Sr and N, but to a lesser extent than in Xe. Diamond particles are more abundant than those of presolar graphite and SiC particles. The sizes of these particles are $\sim$0.001-0.003μm and smaller than those of graphite or SiC (~1μm) particles.

A variety of elemental isotopic measurements of single SiC grains (C, N, Ne, Mg, Si, Ca, Ti) and collections of SiC grains (Ne, Sr, Ba, Kr, Xe, Nd, Sm) have been carried out. There is a variation in the observed $^{12}C/^{13}C$ isotopic ratios. The observed ratio is smaller than the solar value and varies roughly between 40 to 70. This can be seen from Fig. 5.9. SiC can exist in two forms: namely α-SiC and β-SiC. Both the types are present in meteorites. These presolar SiC grains constitute the majority (99%) of the particles present in meteorites. The rest 1% of the SiC particles, is called 'X-type'. These particles show excess of ^{28}Si.

SiC grains and a graphite grain has been detected in the Murchison meteorite. These grains have low $^{12}C/^{13}C$(4-9) and $^{14}N/^{15}N$(5-20) ratios and large excess of ^{30}Si ($^{30}Si/^{28}Si$ range to 2.1 times the solar value) and high $^{26}Al/^{27}Al$ ratio.

Spherical graphite grains show the $^{12}C/^{13}C$ ratio to differ by 3 orders of magnitude(~2-6000) as compared to the solar ratio of 89. The analysis of noble gases in these particles show that they are carriers of Ne-E. The isotopic studies of ^{18}O, ^{28}Si, ^{44}Ca, ^{49}Ti, ^{41}K etc. have been carried out.

Figure 5.8 shows a section of a spherical graphite grain from the Murchison meteorite. It shows clearly the presence of a TiC at its center. Inclusions such as MoC and ZrC are also seen from some graphite grains. These refractive carbides must have acted as a nucleus for the graphite to condense. The sizes of the refractive carbides at the nucleus of graphite grain is small, ~0.001-0.05μm in diameter. There are also some graphite grains where the carbides are spread over the grain indicating that they got into the grain as graphite condensation was taking place.

Presolar oxide grains have also been identified in meteorites. The oxide grains are mostly Al_2O_3 (Corundum) and some of the type $MgAl_2O_4$ (Spinal). The average size of the grains is around 1μm. Oxide grains are found to be less abundant compared to carbon-rich grain present in meteorites.

Among presolar nitride particles, silicon nitride (Si_3N_4) has been identified in meteorites. Si_3N_4 appears to have two components, α-Si_3N_4 and β-Si_3N_4. They are characterized by a large ^{15}N and ^{26}Mg excess. The silicon nitride particles are much less abundant than presolar SiC particles. Table 3.8 gives a list of all the presolar grains detected in meteorites. It can be seen from Table 3.8 that the sizes of the presolar grains in meteorites are generally larger than the sizes of interstellar dust particles. But the sizes of the graphite particles are much larger. This appears to indicate that

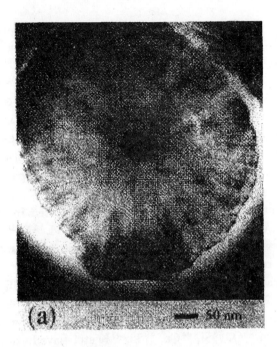

Fig. 5.8 A transmission electron micrograph of a thin section of a presolar graphite grain from the Murchison meteorite. The long nucleus at its center is TiC crystal of $0.07\mu m$ in size. This must have acted as a nucleation center for graphite to condense (Bernatowicz, T.J. 1997, In *From Stardust to Planetesimals*, Eds. Y.J. Pendleton and A.G.G.M. Tielens, ASP Conf. Ser., Vol.122, p.227: By the kind permission of the Astronomical Society of the Pacific Conference Series).

due to some process smaller size grains are selectively destroyed during the formation of the meteoritic parent body.

Most of the presolar grains discussed so far refers to particles condensing out of high temperature material($T \geq 1330K$).

PAHs have been detected in samples of the Allende and Murchison meteorite from the studies of laser desorption mass spectroscopy. The mass spectra show several peaks in the $m/z = 100$ to $m/z = 350$ amu corresponding to PAHs and their alkylated derivatives. Fullerenes have also been detected in Allende and Murchison meteorites.

The abundances of various presolar grains vary from ~ 400ppm for nanodiamonds to a few ppm for graphite and SiC. For other types of particles, it is ≤ 1ppm. Many of the meteorites show enhanced D/H and $^{15}N/^{14}N$ ratios in bulk samples. The primary carrier of these anomalous ratios are the organic compounds. The largest ratio of D/H seen in meteorites is found to be much smaller than D/H ratio seen from cold molecular clouds.

This implies that the original material has been modified due to chemical processes within the solar system.

5.3.3 Time of Formation

The study of meteorites has given important information regarding the time of formation of various bodies in the solar system. i.e. an estimate for the age of the solar system. The meteorites contain several radioactive elements. The formation age of a meteorite can therefore be determined from the natural radioactive decay of these elements. The decay schemes that have been used for dating a meteorite are, ^{87}Rb-Sr, ^{147}Sm-Nd, ^{40}K-Ar, ^{232}Th-Pb and the two ^{238,235}U-Pb systems. The decay constants for these systems are well known. For the calculation of ages, it is necessary to correct for initial or nonradiogenic isotopes which is also available. When a large number of samples are available a plot of parent vs daughter nucleids each normalized to a radiogeneric daughter-element nucleide gives a straight line. The slope of the line is a function of the age of the sample. Hence age can be determined. The above procedure assumes that all the systems are coeval, the initial isotopic composition of the daughter element are the same for all the samples etc. Large number of meteorites have been studied for their age and they give nearly the same value $\sim$4.5Gyr.

The dating of Allende meteorite is of particular interest. This is due to the fact that photographic studies of chondrules and the inclusions in Allende meteorite have shown the presence of highly refractory minerals rich in Ca, Ti and Al. This is similar to the expected composition of the condensates that could form in the early stages of the solar nebula. This implies that the material present in Allende could be the oldest material to be found in meteorites. Therefore extensive dating has been carried out on Allende meteorite. The derived age is 4.559±0.004Gyr. Therefore the derived age of meteorites by several methods and from a sample of a large number of meteorites is $\sim$4.55Gyr. Therefore the age of the solar system derived from meteoritic studies $\sim$4.55Gyr.

Since meteorites are the oldest objects in the solar system they are closer to the Sun in chemical composition than any other known natural material. As meteorites can be analysed in the laboratory it has the greatest advantage and hence accurate measurements of the chemical composition can be carried out. Chemical composition of various elements have been derived from the studies of carbonaceous chondrites. These results supplement the values derived from spectroscopic analysis of solar photosphere for the standard composition of the solar system material.

5.4 Extraterrestrial Origin

Most of the meteorites are believed to be the fragment of asteroids. This comes from chemical and structural properties of meteorites.

The meteorites which have a wide variation in their mineral composition can be used to limit the probable parent bodies. The different kinds of minerals seen in meteorites arise mainly from different degree of thermal evolution, where alteration in chemical composition can take place. But carbonaceous chondrites are not subjected to high temperature phase as they are more primitive than other meteorites. Therefore ordinary meteorites are subjected to severe metamorphism. This implies that stony-iron, iron meteorites and achondrites have gone through the process of complete melting and then recrystalized. In view of these considerations comets cannot be the source of these meteorites as they remained at low temperature ($\leq 300K$) from the time of their formation. However it is quite possible that they are the source of primitive meteorites. Hence most of the meteorites are believed to come from asteroids. This is also consistent with the fact that the wide variety of asteroids type seen, cover the observed meteoritic types. Therefore it is possible that iron meteorites, stony-irons and achondrites could come from different parts of asteroids which are exposed to different temperatures. For such a scenario to be feasible, some heat source is required to melt the asteroid which could come from the short lived ($\sim$million years or so) radioactive elements such as ^{26}Al. This heat is sufficient to melt an asteroid of a few kilometre in size. If asteroids are the source of meteorites then it has to crumble. There is some evidence for this.

The Internal structure of many meteorites show evidence for the presence of fragmentation of rocks. This indicate that some collision event must have taken place. This could arise from the voilent impacts on the surface of the parent body.

However dynamical considerations pose some problems. This is because meteorites in the asteroid belt (in circular orbits) has to be transported to Earth. It is quite possible that due to collisions in the asteroid belt and subsequent gravitational perturbation a group of asteroids could have been created whose orbit come close to the Earth. Infact there is a group called Apollo-Amor which come close to the Earth. This could be source of meteorites.

5.4.1 Presolar Grains

The source of some of the presolar diamond particles could be supernovae. This comes from the fact that Xe isotopes, ^{134}Xe, ^{136}Xe and ^{124}Xe, ^{126}Xe present in diamonds are produced by the r- and p-processes respectively, in supernova (Appendix C). The observed isotopic ratio, ^{12}C/^{13}C close to the solar value possibly indicate that some contribution may be from red giant stars.

The elemental isotopic measurements of single SiC grains and collections of SiC grains have given valuable information about the possible source of these grains. The presence of isotopic signatures of s-process elements in SiC grains suggest that around 99% could arise from low mass AGB stars. Supportive evidence comes from the fact that the isotopic ratio ^{12}C/^{13}C in both the cases are very similar and in addition SiC has been seen in the envelopes of AGB stars. This can clearly be seen from Fig. 5.9. The rest of 1% SiC grains called X-grains appears to have come from supernova ejecta. This is inferred from the presence of large amounts of ^{28}Si and ^{44}Ca produced from the decay of short lived ^{44}Ti, which are produced in the interior of massive stars and ^{15}N produced in explosive nucleosynthesis. Therefore for the grains to contain ^{28}Si and ^{44}Ca, mixing must take place between the stellar interior material with the atmosphere material before the condensation of the grains can take place.

The observed isotopic ratios of ^{12}C/^{13}C, ^{14}N/^{15}N and ^{30}Si/^{28}Si in graphite and SiC grains from the Murchison meteorite are similar to the predicted ratios from the ejecta of a nova. Therefore the particles detected in Murchison meteorite could be attributed to grains from a nova ejecta.

The isotopic abundances of carbon and other elements such as ^{18}O, ^{28}Si, ^{44}Ca, ^{49}Ti, ^{41}K etc. in presolar spherical graphite particles has indicated the complexity of the grain which can come about from contributions from various stellar sources such as Wolf-Rayet stars, supernova, AGB atmospheres etc. This can be seen from the distribution in ^{12}C/^{13}C ratios as shown in Fig. 5.9. The source for the lowest ratios of ^{12}C/^{13}C may be from novae as seen from Murchison meteorite.

The spherical graphite grain with TiC at its center seen from the Murchison meteorite is shown in Fig. 5.8. This clearly indicate that it is a presolar graphite grain. The isotopic ratios indicate that the grain must have originated from the carbon star. This is consistent with the

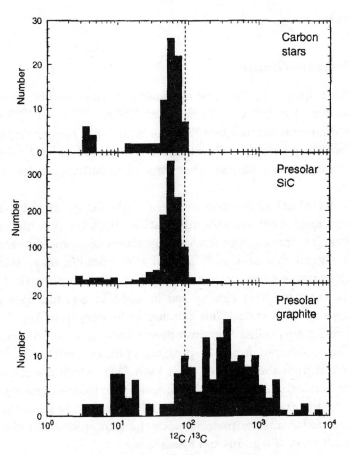

Fig. 5.9 Shows the distribution of the isotopic ratio $^{12}C/^{13}C$ in the atmosphere of carbon stars and presolar SiC and graphite grains seen in meteorites. The dashed line represent the mean solar value (89). The source of presolar SiC grains is consistent with that of carbon stars. The presolar graphite grains appears to come from various sources (Whittet, D.C.B. 2003, *Dust in the Galactic Environment*, Institute of Physics Publishing, Bristol).

condensates expected from such stars.

FeS (troilite) which is a common mineral present in meteorites is also detected from Herbig Ae/Be stars. This indicates the FeS grains in meteorites to be of presolar origin (See Fig. 5.4).

Some oxide grains are found to have depletion of ^{18}O by several factors while other grains have excess of both ^{18}O and ^{17}O compared to the solar system value. These presolar grains are likely to have been produced in oxygen-rich stellar environment. To produce the observed isotopic abundances, there should be a mixing of material from deep interior with the outer surface material. This could possibly happen during the dredge-up episode phase of the star.

It is likely that nitride particles could originate in supernovae.

The variety of materials seen from meteorites clearly show that they are assemblages of debris from different stars born in different molecular clouds at different times as well as due to heating and cooling process during the formation of the planetary system. The presence of nuclei with $\tau < 10^7$ yrs, (i.e. ^{41}Ca, ^{26}Al, ^{60}Fe, and ^{53}Mn) clearly indicate the late stage addition of materials to the solar system. Therefore, meteorites contain reliable record of the first several million years of solar system history and also of individual stellar contributions to the early Sun.

References

IDPs

Bradley, J.P., Sandford, S.A. and Walker, R.M. 1988, In *Meteorites and The Early Solar System*, Eds. J.F. Kerridge and M.S. Matthews, Univ of Arizona Press, Tucson, p.861.

Brownlee, D.E. 1978, In *Cosmic Dust*, Ed. J.A.M. McDonnell, John Wiley and Sons, p.295.

Hanner and Bradley 2004, In *Comets II*, Eds. M. Festou et al., University of Arizona Press, Tucson.

Jessberger, E.K. et al. 2001, In *Interplanetary Dust*, Eds. E. Grun, Bo.A.S. Gustafson, S. Dermott and H. Fechtig, Springer-Verlag, p.253.

Messenger, S. 2000, In *Astrochemistry: From Molecular Clouds to Planetary Systems*, IAU Symposium No. 197, Eds. Y.C. Minh and E.F. van Dishoeck, ASP Publishers, p.527.

Messenger, S., Keller, L.P., Stadermann, F.J., Walker, R.M. and Zinner, E. 2003, Science, **300**, 105.

Sandford, S.A. 1987, Fundamental Cosmic Phys., **12**, 1.

Aromatic and Aliphatic

Maechling, C.R., Zare, R.N., Sucan, P.D. and Walker, R.M. 1993, Science, **262**, 721.

Meteorites

Amari, S.S., Gao, X., Nittler, L.R., Zinner, E., Jose, J., Hernanz, M.and Lewis, R.S. 2001, Astrophys. J., **551**, 1065.

Bradley, J.P., Snow, T.P., Brownlee, D.E. and Hanner, M.S. 1999, In *Solid Interstellar Matter: The ISO Revolution*, Eds. L. d'Hendecourt, C. Joblin and A. Jones, Springer-Verlag, p.297.

Kerridge, J.F. and Matthews, M.S. (Eds.) 1988, *Meteorites and The Early Solar System*, University of Arizona Press, Tucson.

Presolar grains.

Lewis, R.S., Tang, M., Wacker, J.F., Anders, E. and Steel, E. 1987, Nature, **326**, 160.

Bernatowicz, T.J. 1997, In *From Stardust to Planetesimals*, Eds. Y.J. Pendleton and A.G.G.M. Tielens, ASP Conference Series, Vol.122, p.227.

Dai, Z.R., Bradley, J.P., Joswiak, D.J., Brownlee, D.E., Hills, H.G.M. and Genge, M.J. 2002, Nature, **418**, 157.

Chapter 6

Circumstellar Dust

6.1 Introduction

Asymptotic Giant Branch Stars (AGB) are constant suppliers of dust to interstellar medium. Supernovae may also contribute dust to interstellar medium. It is interesting that circumsteller shells of AGB stars can produce either carbonaceous or silicate type of dust particles depending upon the number of thermal pulse episodes which is a function of large scale mass-loss from the star. Infact the time scale of evolution of Post-AGB stars are so rapid that both, carbonaceous and silicate type of dust particles produced at two different episodes in succession can also be present. The dust grains from AGB stars and Novae and the explosive nucleosynthesis products from Supernovae have left their imprints in meteorites and IDPs generally called presolar grains. AGB stars are also suppliers of highly complex molecules, PAHs etc. to interstellar medium. Since the same processes must be operating in similar objects in the universe, dust is universal. These aspects will be discussed in this chapter.

6.2 AGB Stars

Asymptotic Giant Branch stars are intermediate mass stars of around 1 to $8M_\odot$, which have gone through hydrogen and helium burning stages (Appendix A). They are relatively cooler objects with high luminosity and larger radius. The typical parameters for AGB stars are, $T_{eff} \leq 3000K$, $L \sim 10^3\text{-}10^4 L_\odot$ and $R \sim 200\text{-}400 R_\odot$. The mass-loss rate range from 10^{-7}-$10^{-4} M_\odot$/yr and outflow velocities $\sim$10-15km/sec. The core of the star is surrounded by helium followed by an extended H-rich envelope. At the base of the hydrogen-rich envelope, helium can be formed through hydrogen fusion. This leads to helium flash. This is the Thermally Pulsing AGB phase. The most important aspect of AGB stars lie in the nucleosynthesis which occurs during their evolution. During thermal pulse phase convection in the mantle enter deeper layers. This leads to mixing of heavier elements namely, carbon, nitrogen and oxygen, from the stellar interior with the hydrogen-rich envelope of the star. This changes the elemental composition in the atmosphere of the star. The most important element of interest is carbon. Since carbon is produced in larger numbers than oxygen in the interior it is possible that due to repeated thermal pulses (dredge-up) the number of carbon atoms in the atmosphere could exceed that of oxygen atoms. Therefore stars in the AGB phase can lead to two differnt kinds of stars. Those with more oxygen than carbon (i.e C/O < 1) are generally called Oxygen-rich stars (Miras) and the other with more carbon than oxygen (i.e C/O > 1) are called Carbon stars. The stars which have equal number of O and C atoms (i.e C/O =1) are generally called S-type stars. The temperature in the atmosphere of AGB stars is low, hence molecules can form. If C/O > 1 i.e.carbon stars, mostly carbon molecules are formed. If C/O < 1 i.e.Miras, molecules with oxygen are formed. The characteristic spectra of carbon stars is the presence of C_2, CN and CH bands, while Miras have strong TiO bands and S stars have ZrO bands. Infact these bands are used in the spectral classification of these objects (Appendix A).

The Post-AGB stars covers a wider range of T_{eff} and roughly a constant luminosity, $\sim 10^4 L_\odot$. The evolutionary time scale of these stars are very fast, $\sim 10^3$-10^4yrs. Therefore the fast evolutionary phase is characterized by a rapid change in the properties of these objects. Hence a wide variety of chemical signatures coming out of circumstellar mineralogy is expected from these stars. Infact a wide variation in the properties of these stars have been seen.

6.3 Mass Loss from Stars

It has been known from earlier observations that stars with spectra later than spectral type M5 and of luminosity classification I, II, or III are found to be variable. It could be regular or semiregular variation. The variability is detected both in the visual and infrared wavelengths. These stars generally lie in the region expected of AGB stars. The period of variability of these stars could be anywhere between 100 to 500 days or so. The existence of extended circumstellar shells and mass loss from these stars comes from various considerations. The early spectroscopic observations in the visual range were the first to show the presence of circumstellar absorption lines which are displaced towards the violet region of the spectrum, $\sim$ 10km/sec. There are other ways of detecting circumstellar shells which are of more recent origin. The infrared observations clearly showed the presence of thermal emission due to dust particles that form in the outflow. Stars losing mass are also maser sources produced at farther out in the shell. The shells also produce emission lines of rotational transitions of various molecules. The IRAS colour-colour plot (log (F_{12}/F_{25}) vs log (F_{25}/F_{60})) is also a good indicator. Here 12, 25 and 60μm refer to wavelength of the bands. In this diagram, the evolved sources are found to lie along a band of increasing log (F_{12}/F_{25}) of IRAS data which represent increasing rates of mass loss for M and C stars. All these and other types of observations could be used for estimating the mass loss rates, outflow velocities in the shell, extent of the shell etc. For example the mass-loss rate for Sun $\sim 10^{-14} M_\odot$/yr. Typical values for red giants and supergiants of spectral type M2III and M2I $\sim 10^{-8}$ and 10^{-6} $M_\odot$/yr respectively. Several empirical laws for mass loss have been deduced from the observations which gives a correlation between the observed mass loss rates and the combination of parameters for luminosity (L), radius (R) and mass (M) for giants and supergiants. A typical combination of M, L and R is given by the relation

$$\frac{dM}{dt} \propto \frac{L}{gR} \tag{6.1}$$

where L and R are the luminosity and radius of the star and $g = GM/R^2$ is the gravitational acceleration at the surface of the star.

$$\text{i.e.} \quad \frac{dM}{dt} = k\frac{L}{gR} \tag{6.2}$$

where k is the constant of proportionality. k has been estimated from the observed mass-loss rates from red giant stars, which gives, k$\sim 4 \times 10^{-13}$ $M_\odot$/yr. The above relation gives mass-loss rate $\sim 10^{-8}$-10^{-9} $M_\odot$/yr for a star of MIII type.

6.4 Theoretical Considerations

6.4.1 *Dust Formation*

It has been recognised that the circumstellar shells of stars provide the ideal conditions for the formation of dust particles which are the continuous source of dust grains to the interstellar medium. After they are formed they have to be expelled from the circumstellar shell in some form of wind. The first mechanism to be considered was within the framework of stationary outflow as in the case of Solar wind. However it is rather difficult to produce the observed high mass loss rates with models based on this mechanism. The next logical step is to include dust formation along with stationary flow. It is well known that dust particles can scatter and absorb the incident radiation and hence can pick up momentum from the radiation field. The dust grains and the gas is coupled through grain-gas collisions. How effective the transfer of momentum to the gas which drives the gas is dependent upon the optical properties of the dust particles. Detailed model calculations which have considered the nucleation and growth of grains, the momentum transfer from dust to gas based on the optical properties of dust grains for conditions expected in AGB stars, have shown that steady winds driven by dust alone is highly unlikely as it is rather difficult to produce the observed mass loss rates. However the dust formation in a pulsating extended atmosphere which drives the stellar wind leading to enhance mass loss rate is most attractive mechanism. Pulsation has several inherent advantages. It can give rise to high densities and lower temperatures where dust can form. In addition, the standing or travelling waves produced by pulsation can lead to shock waves passing through a decreasing density shell. All these can help in pushing the material out. The formation of grain is basically through a non-equilibrium process and this condition is met in a pulsating atmosphere. The dust formation can also take place closer to the star. Detailed model calculations have shown that it is possible to get regions where the pressure exceeds the saturation pressure by several orders of magnitude. These are the regions where nucleation of dust particles should be rapid and move with small outward velocity. The coupling of the gas and the dust leads to net outflow. The mass loss rates are found to be strongly dependent on stellar parameters namely, L, M and chemical composition. For example, in general for a given M and L, the mass loss rates are lower for lower Z (abundances of elements other than H and He) values. The general conclusion that comes out of theoretical studies is that pulsation along with dust formation in an extended atmosphere drives the stellar winds is a highly attractive mechanism.

6.4.2 Condensation of Dust

The condensation and growth of dust particles in the circumstellar shells of stars is complicated as it involves full description of chemistry describing the passage of the gas to solid phase under physical conditions far from equilibrium. In addition the formation of dust itself influences the flow. The formation of dust particles is a two step process. Firstly, the formation of a condensation seed, followed by growth by addition of suitable molecules leading to small solid particles. The molecule CO plays a dominant role in the type of grain that can form. This is because CO is a strongly bound molecule and cannot easily be broken in simple gas-phase reactions. It is also very volatile and cannot condense into solids. It will also not attach to particles. Therefore the element C or O, whichever is less abundant is completely locked up in CO. Hence that element is not available to form solids. Therefore C/O ratio in the atmosphere of stars plays a dominant role in deciding the type of grains that could form.

The nature of dust grains that could possibly condense in the circumstellar shell of stars is of great interest. Once the nuclei are formed they have to survive and grow to get to reasonable grain sizes. There are a large number of studies relating to these aspects under equilibrium and non-equilibrium cases. However a realistic description of the dust formation in circumstellar shells involves the study that include hydrodynamics, thermodynamics, chemistry, dust formation, growth and evaporation of grains, radiative transfer in the shell and so on. This is a highly complex problem. However it is instructive to start the discussion based on chemical equilibrium situations. For a general study it can give a good idea of the expected results. This approach has been successful in the past. The type of material that condenses out depend on temperature, pressure and elemental composition and can be predicted from chemical equilibrium calculations. The type of condensates depend upon C/O ratio.

6.4.2.1 $C/O < 1$ (Oxygen-rich stars)

The condensation temperature of major compounds as a function of pressure for an atmosphere with solar composition is shown in Fig. 6.1. As can be seen from the figure, condensation temperature generally decrease

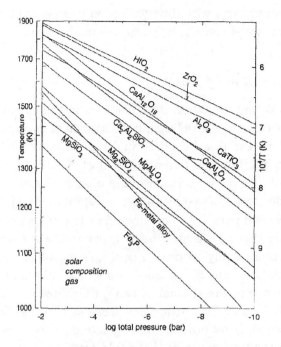

Fig. 6.1 Condensation temperature of major elements plotted as a function of total pressure for a solar composition gas for oxygen-rich case (C/O=0.48) (Lodders, K and Fegley, B. 1999, In *Asymptotic Giant Branch Stars*, Eds. T. Le Bertre, A. Lebre and C. Waelkens, IAU Symposium No.191, p.279: By the kind permission of the Astronomical Society of the Pacific Conference Series).

with the decrease of total pressure. The production of solids depend upon the most abundant species namely, Fe, Mg, SiO and H_2O. At temperatures greater than 1500K, the stable solids are the refractory oxides Al_2O_3 (Corundum) and $CaTiO_3$ (Perovskite). The initial solids can then easily transform into other products. For example, Al_2O_3 can react with Ca (gas) to form $CaAl_{12}O_{19}$ (Hibonite). This then leads to grossite and finally to gehlenite (Table 6.1). Similarly, Al_2O_3 can react with Mg(gas) and H_2O (gas) to produce spinel($MgAl_2O_4$). SiO leads to SiO clusters. These SiO clusters take on Mg leading to $MgSiO_3$ and Mg_2SiO_4. At lower temperatures, T~700K, most of the elements would have condensed into solid form. For still lower temperatures, Fe is in FeO form and finally at temperature ~200K, H_2O condenses. Therefore in an oxidizing environment refractory oxides and silicates are the condensates.

Table 6.1 Major elemental condensates expected in M and C stars*.

Element	Abundance§	M-stars Formula	M-stars Mineral Name	C-stars Formula	C-stars Mineral Name
O	2.09×10^7	oxides and silicates			silicates
C	1.00×10^7	–	–	TiC	titanium carbide M
				C	graphite CS,M
				SiC	silicon carbide CS,M
N	2.63×10^6	–	–	TiN	osbornite
				AlN	aluminum nitride
Mg	1.02×10^6	$MgAl_2O_4$	spinel M	MgS	niningerite CS
		Mg_2SiO_4	forsterite CS	$MgAl_2O_4$	spinel M
		$MgSiO_3$	enstatite CS	Mg_2SiO_4	forsterite CS
				$MgSiO_3$	enstatite CS
Si	1.00×10^6	$Ca_2Al_2SiO_4$	gehlenite	SiC	silicon carbide CS,M
		Mg_2SiO_4	forsterite CS	FeSi	iron silicide
		$MgSiO_3$	enstatite CS	Mg_2SiO_4	forsterite CS
				$MgSiO_3$	enstatite CS
Fe	8.91×10^5	Fe	iron metal	FeSi	iron silicide
		$(Fe,Ni)_3P$	schreibersite	Fe	iron metal
		FeS	troilite	FeS	troilite
S	4.47×10^5	FeS	troilite	CaS	oldhamite
				MgS	niningerite CS
				FeS	troilite
Al	8.51×10^4	Al_2O_3	corundum CS,M	AlN	aluminum nitride
		$CaAl_{12}O_{19}$	hibonite	Al_2O_3	corundum CS,M
		$CaAl_4O_7$	grossite	$MgAl_2O_4$	spinel M
		$Ca_2Al_2SiO_7$	gehlenite	$CaAl_2Si_2O_8$	anorthite
		$MgAl_2O_4$	spinel M		
		$CaAl_2Si_2O_8$	anorthite		
Ca	6.46×10^4	$CaTiO_3$	perovskite	CaS	oldhamite
		$CaAl_{12}O_{19}$	hibonite	$CaAl_2Si_2O_8$	anorthite
		$CaAl_4O_7$	grossite		
		$Ca_2Al_2SiO_7$	gehlenite		
		$CaAl_2Si_2O_8$	anorthite		
Na	5.75×10^4	$NaAlSi_3O_8$	albite	$NaAlSi_3O_8$	albite
Ni	5.01×10^4	FeNi	kamacite & taenite	FeNi	kamacite & taenite
		$(Fe,Ni)_3P$	schreibersite	$(Fe,Ni)_3P$	schreibersite
Cr	1.35×10^4	Cr in FeNi	alloy	$FeCr_2S_4$	daubréelite
Mn	9.33×10^3	Mn_2SiO_4	rhodonite in olivine	$(Mn,Fe)S$	alabandite
P	8.13×10^3	$(Fe,Ni)_3P$	schreibersite	$(Fe,Ni)_3P$	schreibersite
Cl	5.25×10^3	$Na_4[AlSiO_4]_3Cl$	sodalite	NaCl (?)	halite (?)
K	3.72×10^3	$KAlSi_3O_8$	orthoclase	$KAlSi_3O_8$	orthoclase
Ti	2.40×10^3	$CaTiO_3$	perovskite	TiC	titanium carbide M
				TiN	osbornite

* condensates are listed in order of appearance for each element.
§ solar abundances on a scale Si = 1×10^6 atoms (Lodders & Fegley, 1998)
CS: dust observed in circumstellar shells of AGB stars.
M: circumstellar grains found in meteorites.

Lodders, K. and Fegley, B.Jr. 1999, In *Asymptotic Giant Branch Stars*, IAU Symposium No. 191, Eds. T.Le. Berre, A. Lebre and C. Waelkens, p. 279: By the kind permission of the Astronomical Society of the Pacific Conference Series.

6.4.2.2 C/O >1 (Carbon-rich stars)

The condensation temperature of major compounds as a function of pressure is shown in Fig. 6.2. As in the previous case, here again the condensation temperature decreases with decrease in total pressure. At higher temperature, ZrC (Zirconium Oxide) and TiC (Titanium Carbide) are the condensates. However the abundance of ZrC is very much smaller than TiC because of the difference in their elemental abundances (Ti:Zr=10^5:1). When the temperature drops below 1600K, graphite forms.

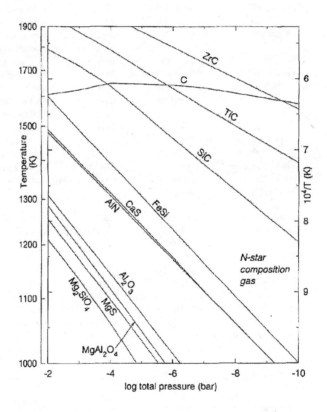

Fig. 6.2 Condensation temperature of major elements plotted as a function of total pressure for a solar composition gas for carbon-rich case (C/O=1.1) (Lodders, K. and Fegley, B. 1999, In *Asymptotic Giant Branch Stars*, Eds. T. Le Bertre, A. Lebre and C. Waelkens, IAU Symposium No.191, p.279: By the kind permission of the Astronomical Society of the Pacific Conference Series).

6.4.2.3 C/O = 1 (S-Stars)

S Stars with C/O≈1, represent the transition from oxygen-rich stars to carbon-rich stars. The non-availability of sufficient quantities of either O or C leads to vastly different condensates compared to those of M stars and C stars. The exact mineral mixture that can form depend critically on C and O abundances and temperature. Chemical equilibrium calculations have been carried out for values of C/O covering the region, slightly less than 1 to slightly more than 1. These results indicate that most likely condensates are FeSi, metallic iron and small quantities of forsterite (Mg_2SiO_4) and SiC. Therefore solid iron and FeSi should be dominant dust specie in the circumstellar shell of S stars.

In carbon stars, acetylene(C_2H_2) is the most abundant molecule. The reaction involving C_2H_2 and C_2H_2 leads finally to Benzene a six membered ring as an initial ring compound. Subsequent reactions build up more complex compounds. The overall effect is to convert acetylene to PAH. Detailed studies have shown that the amount of PAH that can form is sensitive to various quantities and hence its production is limited to a narrow range of temperature (900-1100K). Therefore the homogeneous nucleation of PAH molecules to form a carbon solid is highly unlikely. Hence it is more likely that SiC grains, the seed nuclei, are coated with amorphous carbon containing PAH groups. This material may be the unsaturated form of amorphous carbon. Unlike other condensates, for graphite the condensation temperature is not very sensitive to pressure. Therefore the condensation sequence for TiC, graphite and SiC is pressure sensitive. i.e. a particle can have TiC as the nucleus and graphite over it etc. In both the cases, i.e. for C/O<1 and C/O>1, each condensate can act as a nucleus for other condensates to grow over it. It is quite possible that a final grain will have concentric layers of different kinds of condensates starting from high temperature phase in the innermost regions and with lower temperature condensates moving outwards. Based on chemical equilibrium studies the type of condensates that can condense in C and M type stars are given Table 6.1. Many of the condensates given in Table 6.1 have been seen in circumstellar shells of stars and in meteorites. They are indicated as CS and M respectively in the Table. Some of the condensates have not been seen around cool stars. This could partly be due to the nature of calculations. In addition the type of condensates that is possible also depend on the bulk elemental composition. The C/O ratio also influences the condensation temperature.

6.4.3 Circumstellar Chemistry

The circumstellar shell chemistry is also relevant as it can influence the nature of the dust grains through catalysis of gas-phase molecules on dust grain surfaces. They could also give mantles around bare grains. Besides, highly complex organic molecules including PAHs produced in the circumstellar shell of AGB stars provide bulk of material to interstellar medium. The chemistry in circumstellar shells of carbon stars has been studied extensively based on photochemical model.

The basic idea of photochemical model is that the most dominant molecules CO, C_2H_2 and HCN, produced in AGB stars travel outwards in the circumstellar shell due to wind. The material then reaches the region in the shell where it starts experiencing the interstellar radiation field. This leads to photodissociation of the species producing radicals. Ionization of the species also takes place. This will then initiate chemical reactions. Neutral reactions and ion-molecular reactions are considered. The results of these calculations clearly show that starting from simple molecules, i.e. CO, C_2H_2 and HCN, it is possible to build up in steps, molecules of higher complexity leading finally to highly complex organic molecules such as Cyanopoyynes, HC_3N, HC_5N etc. and carbon chain radicals, C_5H, C_6H etc. In the actual calculation, large number of reactions involving several elements C, N, O, Si, S, Mg and other heavier elements are generally considered. However in any particular situation, only a subset are important The following illustration shows the main mechanism responsible for producing the simplest polyacetylene, HC_3N.

The production of HC_3N can take place via ion-molecule reaction as,

$$h\nu + C_2H_2 \to C_2H_2^+ + e$$
$$C_2H_2^+ + HCN \to H_2C_3N^+ + H$$
$$e + H_2C_3N^+ \to HC_3N + H$$

and via neutral reaction as,

$$h\nu + HCN \to CN + H$$
$$CN + C_2H_2 \to HC_3N + H.$$

Out of the above two processes, the production of HC_3N through neutral reactions dominate, as CN abundance is around 10^2 to 10^3 times larger than that of $C_2H_2^+$.

Dissociative recombination of molecular ions can produce isomeric as well as normal forms. For example,

$$e + HCNH^+ \to HCN \tag{6.3}$$

can give HCN as well as H + CN. The ring molecule C_3H_2 can be produced in the dissociative recombination of $C_3H_3^+$.

These results show that various types of molecules including highly complex molecules are expected to be produced in the circumstellar shells. These are well supported by the observations of circumstellar shells of carbon stars, such as IRC+10216. The star IRC+10216 with a rich carbon chemistry is a very well studied object. This bright star was first detected in the $2\mu m$ survey and is a long period AGB star with a period of 649 days and with a mass loss of few times $10^{-5} M_\odot$/yr. High resolution radio line observations have indicated the expansion velocity in the shell of 14 km/sec. The relative abundances of CO, C_2H_2 and HCN with respect to total hydrogen, derived from observations is 4×10^{-4}, 4×10^{-5} and 2×10^{-5} respectively. These values refer to the region around photodissociative zone.

The observations carried out at optical, infrared and radio wavelength regions have made it possible to detect a large number of molecules in IRC+10216. The species detected from IRC+10216 is given in Table 6.2. As can be seen from Table 6.2 more than 50 species have been detected. The species given in Table 6.2 are arranged in certain groups(columns) and listed according to their measured abundances. The saturated molecules are given in 3rd column. This is followed by the products associated with the photodissociation and radical chemistry associated with C_2H_2 and HCN. The resulting isomers are given in the next column. In the last two columns S and Si-bearing molecules are given. As can be seen large number of unsaturated carbon and acetylene chains are present. Metal- and Si-bearing species are also present. The molecules given in Table 6.2, as well as the observed spatial distribution of molecules support the photochemical model.

The high resolution spectra of IRC+10216 has been obtained with ISO in the wavelength region 2-200μm. They show a large number of well resolved rotational lines arising out of various molecules. The rotational lines of C_2H_2, HCN etc are present in the $3\mu m$ region and rotational lines of acetylene in the $14\mu m$ region. The region $\lambda > 75\mu m$ is filled with rotational lines of HCN.

From the profile fit to the observations, the derived relative abundance of HCN with respect to hydrogen is about 3×10^{-5}, which is consistent with the value derived from radio observations.

Table 6.2   Species detected in IRC + 10216.

10^{-3}			CO			
10^{-4}			$C_2H_2^*$			
	C			HCN		
10^{-5}			C_2H	CN		
		CH_4^*	C_4H		SiS	CS
10^{-6}			C_3^*	HC_3N		
				C_3N		
				HC_5N	SiC_2	AlCl
10^{-7}		SiH_4^*	C_6H	HNC	SiO	AlF
		NH_3	C_5^*	HC_7N	SiN	NaCN
			cC_3H_2	HC_9N	SiC	MgNC
10^{-8}			lC_3H	H_2C_4	$HC_{11}N$	C_2S	CP
			C_5H	HC_2N	CH_3CN	C_3S	NaCl
				H_2C_3			MgCN
10^{-9}			cC_3H			SiC_4	H_2S	KCl
	HCO^+					
10^{-10}						

[1] C_2 has been detected but its abundance has not been determined.
[2] The species marked with an asterisk have been detected only in the infrared. Glassgold, A.E. 1999, In *Asymptotic Giant Branch Stars*, IAU Symposium No. 191, Eds. T. Le Berre, A. Lebre and C. Waelkens, p.337: By the kind permission of the Astronomical Society of the Pacific Conference Series.

6.5 Observational Results

6.5.1 *Carbon-rich Stars*

The possible presence of carbon particles in the carbon star R Coronea Borealis (RCrB) was long suspected even in 1930's which came from the observation that this star exhibited a large variation in brightness of upto 9 magnitudes without any periodicity. During these events the optical spectrum of the star does not change indicating that the variation is not caused by changes in the star's photosphere. It is more likely to be caused by some obscuring material entering the line of sight of the star. This led to the suggestion that graphite particles formed in the atmosphere of the star cause the dimming of the light.

Many carbon rich stars show strong emission band at 11.2μm. This has been assigned to Si-C vibrations of silicon carbide grains condensed in the stellar wind. SiC emission at 11.2μm is the only spectral signature of dust that is commonly observed from carbon-rich stars. Silicon carbide occurs mainly in two types. α-SiC which has hexagonal structure and β-SiC which has a cubic structure. The two types of SiC have slightly different infrared spectra. Observations of SiC emission seen from carbon stars is of β-SiC type. Despite the fact that SiC is commonly seen from carbon stars, SiC has not been seen in absorption in the interstellar medium. However SiC has been seen as inclusions in meteorites and is presolar in origin. Both α- and β type SiC grain have been seen from primitive meteorites. The other dust related feature present are the absorption features at 3.4 and 6.2μm. The 6.2μm feature appears to be the AUIB feature at this wavelength seen in absorption. It is not clear whether it arises from the interstellar dust or from the dusty material around Wc stars in which it is observed.

A broad and strong feature around 21μm is an important component in the infrared spectra of several post-AGB stars. The identification of the 21μm feature has been a problem for a long time. Several possibilities were suggested for their carriers such as hydrogenated amorphous carbon, nano-diamonds etc. But none of the suggested ones were found to be satisfactory from a comparison with the laboratory spectra. Another suggestion has been made, Titanium carbide (TiC) nono-crystal clusters, based on the excellent match between the laboratory spectra and the observed feature. Possible support comes from the observation that TiC has been seen in presolar graphite grains in meteorites. But this suggestion faces the difficulty that the conditions for its formation in the ejecta of carbon-rich evolved stars may not be met. In addition because of low abundance of Ti, it may be difficulty to produce a strong feature.

The mid infrared observations in the region 6 to 20μm of the carbon star AFGL 5625 showed a broad absorption feature around 10μm. This could be fitted with a combination of SiC and silicate absorption. The full ISO SWS spectra (2-45μm) of the carbon star AFGL5625 showed an additional broad emission around 30μm. This feature was first detected in the far-infrared spectra of C-rich stars and planetary nebulae, CW Leo, IC 418 and NGC 6572, in 1980's. This 30μm feature is a common property of carbon-rich evolved stars. It has been seen in carbon-rich AGB stars, Post-AGB stars and Planetary nebulae. It is seen with varying band shapes and feature to continuum ratios. This feature is generally attributed to Magensium Sulphide (MgS). This suggestion comes from the fact that the laboratory studies show the presence of a resonance line of MgS at this wavelength. This is also consistent with the fact that MgS can easily condense in these objects. The calculated profile of 30μm feature based on MgS grains can successfully explain the observed feature. The elemental abundances of Mg and S are found to be consistent with MgS as the carrier of the feature. Other carriers have also been suggested for the 30μm feature, such as carbon based linear molecules with specific side groups, carbonaceous dust grains with oxygen in the structure, hydrogenated amorphous carbon etc. Because of their chemical instability under oxidizing conditions it is not clear whether MgS can survive through the interstellar space. However the iron-rich sulphides can be expected to occur in interstellar space due to their greater stability. The presence of sulphide bearing GEMS in interplanetary dust particles which are of presolar in origin support this interpretation.

AFGL618 is a C-rich post-AGB star evolving from the AGB to the planetary nebula phase. The central star of AFGL618 has a high temperature $\sim$32000K. For long wavelength region, $\lambda \sim 100\mu$m, the grain emissivity with wavelength can be represented by $\lambda^{-\beta}$, with the value of β in the range 1 to 2. The circumstellar dust emission from carbon rich star AFGL618 can be fitted with a modified Black body B (λ,T) $\lambda^{-\beta}$ for T=60 $\pm$10K and $\beta = 1.9\pm$ 0.2. This suggests that the maximum grain size is less than few tens of μm.

6.5.2 Oxygen-rich Stars

The composition of dust surrounding oxygen-rich AGB stars has been the subject of many studies. The earlier infrared observations had detected silicate bands around 10 and 20 μm which presented first evidence for the silicate nature of dust. This was followed by extensive observations carried out with IRAS satellite which confirmed the silicate nature of the dust particles. These observations also indicated that the dust particles could be amorphous silicate. The feature present around 13μm is often ascribed to Al_2O_3 and emission at 11, 13.1 and 19μm to metal oxides. Support for the identification of aluminium oxide grains comes from the detection of presolar corundum (α-Al_2O_3) grains in primitive meteorites. However accurate profiles derived from ISO-SWS spectra indicate that Al-O vibrations of spinel ($MgAl_2O_4$) may be better. Support for this alternative identification has come laboratory experiments as well as the detection of presolar spinel grains in meteorites. The observations carried out with ISO showed a large number of emission features longward of 20μm and showed the mineralogical diversity present in the circumstellar shell of oxygen-rich stars. The 10 and 20μm silicate emission features present in evolved stars are generally smooth with no structure. However structures should be present if silicate is in crystalline form. Therefore it was a surprise to see crystalline structure of silicate present in the long wavelength region of ISO spectra of AGB stars. The ISO spectra could be used to extract olivine/pyroxene and Mg/Fe ratios which give information on the mineralogy of silicates and also on crystallainity. Figure 6.3 show the continuum subtracted spectra of post-AGB star AFGL4106. The spectrum show the richness of solid state

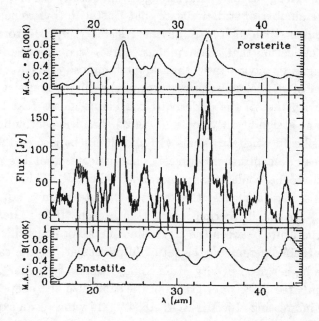

Fig. 6.3 ISO mid-infrared spectra of the post-AGB star AFGL 4106 which show emission lines that can be assigned to crystalline silicate grains. The continuum subtracted spectrum (center) is compared with normalized laboratory-based spectra of forsterite (Mg_2SiO_4) and enstatite ($MgSiO_3$) (Jager, C., Molster, F.J., Dorschner, J., Henning, Th., Mutschke and Waters, L.B.F.M. 1998, Astron. Astrophys., **339**, 904).

features at longer wavelengths. The strongest bands can be attributed to crystalline silicates, both olivines and pyroxenes. Table 6.3 contains a list of the detected features and identifications based on laboratory data. The spectrum is also dominated by amorphous silicates with bands near 10 and 20μm. The features at 43 and 60μm is due to crystalline H_2O ice. The 43μm band is blended with crystalline pyroxene. The wavelength of the band positions suggest that the crystalline silicates are Mg-rich and Fe-poor, incontrast with the amorphous silicates which are believed to be rich in Fe. The Fe/Mg ratio derived from olivine and pyroxene features is quite low, ~0.03-0.05. This is consistent with pure crystalline magnesium silicates $MgSiO_3$ and Mg_2SiO_4. This is in contrast to the ratio of Fe/Mg for amorphous silicates derived from the study of $10\mu m$ feature from circumstellar envelopes and from the interstellar medium, which is ~1.

The relative abundance of olivines to pyroxenes can be determined by measuring the strength of $33\mu m$ pyroxene band and $33.8\mu m$ olivine band and with a knowledge of their band strengths. The study based on ISO spectra of a large number of stars indicate that Enstatite (pyroxene, $MgSiO_3$) is more abundant than forsterite (olivine, Mg_2SiO_4) by a factor of 3 to 4. There is an indication that the luminous sources (red supergiants) have a higher pyroxene to olivine ratio than lower luminosity AGB stars. From a comparison of the colour temperatures of the olivine emission bands in AFGL4106 with that of the underlying continuum shows that the olivine band ($33.8\mu m$) is significantly cooler ($\sim 80K$) than the continuum and the amorphous silicate emission bands (10 and $20\mu m$ bands, $\sim 120K$). This trend has been seen in other sources. This indicates that the crystalline silicates tend to be cooler than amorphous silicates. This difference in temperature seems to suggest that the two silicate grain components are separate and are not in thermal contact with each other. The difference in grain temperature between the crystalline and amorphous silicates could arise due to their different Mg/Fe ratio. In particular, for the case Mg/Fe is small, the higher absorptivity of Fe could give rise to higher grain temperature compared to Fe-poor grains of the same size and shape.

Another interesting object is IRAS 16342-3814 which is an extreme reddened OH/IR star. Maser line emissions from H_2O and OH have been detected from this star. This had indicated the presence of extremelly high velocity outflow in H_2O, ~130km/sec and in OH, ~70km/sec. The ISO spectrum of this star in the wavelength region from 2 to $200\mu m$ shows clearly the amorphous silicate absorption feature around $20\mu m$. In addition, the spectrum shows the characteristic features of crystalline silicates in absorption upto $45\mu m$. At wavelengths longer than $45\mu m$, the circumstellar features present are in emission. This indicates that IRAS 16342-3814 must have recently had an extremelly high mass-loss rate. The features present in the spectra have been identified with silicate materials such as amorphous silicate, forsterite (Mg_2SiO_4), diopside ($MgCaSi_2O_6$) and possibly, clinoenstatite ($MgSiO_3$). In addition to these crystalline silicates, crystalline water-ice has been detected. The spectral features present in IRAS 16342-3814 have also been seen from oxygen-rich evolved stars with

Table 6.3 Solid state bands present in the SWS and LWS spectrum of AFGL4106.

Wave length μm	F_{band} 10^{-14} Wm^{-2}	FWHM μm	FWHM/λ	$\frac{I_{peak}}{I_{cont}}$	Identification
10.8	4.10^2	2.5	0.23	1.8	amorph. silicate
13.6	10.	.5	0.037	1.04	cryst. forsterite
14.2	6.5	.4	0.028	1.05	cryst. enstatite ? instrumental ?
16.1	40.	.7	0.043	1.06	cryst. forsterite
16.8	15.	.6	0.036	1.03	unidentified
17.6	5.10^2	3.	0.17	1.2	amorph. silicate
18.1	75.	1.0	0.055	1.07	cryst. enstatite
19.2	50.	1.0	0.052	1.045	cryst. enstatite and cryst. forsterite
20.6	20.	0.48	0.023	1.035	cryst. enstatite ?
21.5	7.3	0.25	0.012	1.024	cryst. enstatite and cryst. forsterite ?
22.9	25.	0.5	0.021	1.04	unidentified
23.6	30.	0.7	0.030	1.04	cryst. forsterite
24.1	.46	0.09	0.0037	1.004	unidentified
26.2	16.	1.05	0.040	1.02	cryst. forsterite
27.8	44.	1.15	0.041	1.05	cryst. forsterite and cryst. enstatite
32.8	20.	.6	0.018	1.07	cryst. enstatite
33.6	40.	1.0	0.030	1.08	cryst. forsterite
34.0	.96	.12	0.0035	1.016	cryst. enstatite ?
35.8	5.5	.54	0.015	1.03	cryst. enstatite
36.5	2.4	.28	0.0077	1.02	cryst. forsterite
40.4	13.	1.1	0.027	1.05	cryst. enstatite
41.1	1.1	.17	0.0041	1.02	unidentified
43.1	17.	1.2	0.028	1.07	cryst. enstatite and cryst. H$_2$O
47.8	9.	1.9	0.040	1.036	unidentified
61	68.	20.	0.33	1.05	cryst. H$_2$O

Waters, L.B.F.M., Molster, F.J. and Waelkens, C. 1999, From *Solid Interstellar Matter: ISO Revolution*, Eds. L. d'Hendecourt, C. Joblin and A. Jones, Springer-Verlag, p.219: with kind permission of Springer Science and Business Media.

high mass-loss. The ISO spectra of IRAS 16342-3814 indicate that the bulk of the material around this star must be at low temperature. This is also indicated by the presence of 69μm feature of forsterite. There is a feature at 48μm whose identification is not clear but could be FeSi.

It is also seen that not all oxygen-rich AGB stars show crystalline silicates. They tend to appear only when the mass loss rate is high. i.e only in stars with dust shells with color temperature less than 200-300K. The strength and appearance of crystalline silicate bands show a large degree of diversity from source to source.

It is not clear how the crystalline silicates form and why their chemical composition differs from that of the amorphous component. The lack of crystalline silicates in dust shells of low mass loss suggests that the density in the dust forming layers must be high enough for the crystalline silicate to be detected. It is possible that crystalline Mg-rich silicates are the grains that are formed first since their condensation temperature is high. At lower temperatures (< 1000K) amorphous Fe-rich silicates are formed by adsorption of Fe in the lattice of the Mg-rich grains. This can take place at temperature below the glass temperature of silicates and hence the grain becomes amorphous. It is also possible that Fe-rich amorphous mantles grow over Mg-rich grains.

6.5.3 S Stars

Iron Silicide (FeSi) is expected to be the dominant condensate in S type stars. FeSi has two strong features at 32 and 50μm at temperature less than 200K. The two features present at 32 and 47.5μm in the ISO spectrum of AFGL4106 have been attributed to FeSi. Below T<150K, FeSi exhibit four resonance lines at 50.5, 31.4, 29.6 and 22.5μm. The detection of these additional features should give a more definite identification of FeSi in S stars.

6.5.4 UIR Bands in Oxygen-rich Supergiants

The dust formed in an oxygen-rich star is expected to be composed only of oxygen-rich materials namely silicates and oxides. However ISO-LRS spectra of two oxygen-rich M supergiant MZ Cas and AD Per showed an emission feature around 11.5μm, which is attributed to SiC. More observations carried out in the infrared region of several stars have shown the presence of emission feature at 8.6 and 11.3μm, the unidentified infrared bands (AUIBs). These features are superposed on the broad 9.7μm silicate emission feature. While the 8.6μm band is generally seen only along with other members of the ensemble of AUIB features the band at 11.3μm could also be attributed to crystalline olivine. Therefore it was necessary to look for other well known AUIB features for confirming the 11.3μm feature. The ISO-SWS spectrum of the supergiant AD Per did show the presence of bands at 6.2, 7.7 and 8.6 in addition to the 11.3µm band (Fig. 6.4). These bands have also been seen from several other supergiant stars. This showed clearly that aromatic carrier is present around some oxygen-rich M supergiants.

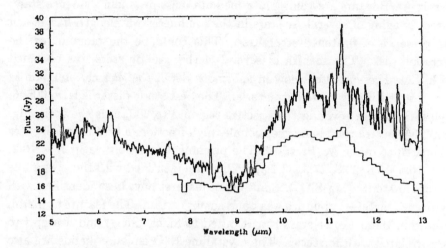

Fig. 6.4 The SWS spectrum of AD Per. The features at 6.2, 7.7, 8.6, 11.3µm are the AUIBs. The broad silicate feature around 10 to 11µm can be seen (Sylvester, R.J. 1999, In *Solid Interstellar Matter: The ISO Revolution*, Eds. L. d'Hendecourt, C. Joblin and A. Jones, Springer-Verlag, p.263: with kind permission of Springer Science and Business Media).

The presence of aromatic carrier around oxygen-rich M supergiants raises the question of its formation as equilibrium condensation theory predicts that all available carbon should be locked up in CO molecules resulting in the formation of only oxygen-rich molecules and particles. The presence of carbon-rich species has been interpreted in terms of a non-equilibrium chemistry. The process involved is that chromospheric UV radiation dissociate CO molecules liberating atomic carbon which enable the formation of carbon-rich species as well as usual silicates. The same UV radiation excites the molecules giving rise to infrared emission bands.

6.5.5 *Stars of Other Type*

Circumstellar emission is not just limited to highly evolved stars. They have also been seen in various kinds of other stars such as young stellar objects (YSOs), pre-main sequence stars such as Herbig Ae/Be stars and T-Tauri stars, main sequence stars such as Vega etc., indicating the presence of dust around these objects.

6.5.5.1 Herbig Ae/Be Stars

Herbig Ae/Be stars are young, intermediate mass pre-main sequence stars. Direct imaging of several Ae stars in the millimeter wavelength has shown the presence of flattened structures. This could be the remnant of the accretion disk. The spectra of several Herbig Ae/Be stars was obtained with ISO. These spectra show in enormous detail the thermal emission of dust from their circumstellar material. The best studied star is HD 100546. The ISO spectra covering the spectral region 2 to 200μm is shown in Fig. 6.5. The spectra reveal an exceptionally high fraction of crystalline silicates compared to other Ae/Be stars. The prominent emission features at 10.2, 11.4, 16.5, 19.8, 23.8, 27.9, 33.7 and 69μm are identified with the crystalline silicate forsterite (Mg_2SiO_4). Some of the features have been identified with pyroxene ($MgSiO_3$), but are weak and so not certain. The feature at 8.2μm seen from these Ae/Be stars is attributed to Silica (SiO_2) and it seems to be correlated with forsterite. The crystalline H_2O emission bands are also present at 43 and 60μm. The characteristic AUIB features are also present. Thermal emission dust modeling has been carried out. Thermal emission from a grain mixture of $[Mg,Fe]SiO_4$, FeO, C, H_2O, forsterite, olivine and Fe gives a good fit to the observed spectra over the entire wavelength region. The deduced mass fractions of amorphous silicate : carbon : metallic iron : forsterite = 0.88 : 0.09 : 0.01 : 0.02. Therefore the major fraction of the dust emission seen from Herbig Ae/Be stars is due to amorphous silicates. The degree of crystallinity of silicates is typically of the order of 5-10%. Even though FeO has been used in the mixture, FeS may be preferred over FeO, as FeS has been identified in Herbig Ae/Be stars (see below). Both FeO and FeS have very similar properties.

The infrared spectra of Herbig Ae/Be stars, HD97048 and Elias 1, in the wavelength region 3 to 3.5 μm, showed the presence of two strong and sharp emission features. The peak wavelength of the features were around 3.43 and 3.53μm. Several observations of these features have shown the presence of detailed structure. This can be seen from Fig. 6.6. These features have also been seen in the infrared spectrum of the Post-AGB star HD4049. There is an indication of the presence of some feature at 3.40 and 3.53μm in the infrared spectrum of supernova SN1987A. As laboratory studies indicate (Sec. 2.5.2), hydrogenated nanodiamond particles gives rise

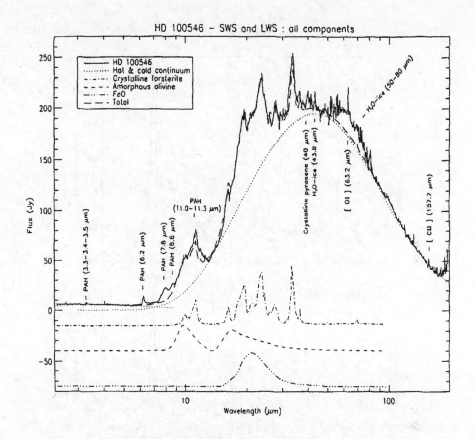

Fig. 6.5 The full SWS-LWS spectrum of HD100546 shows a large variety of emission features. A good match to the whole spectrum requires components of crystalline and amorphous silicates, FeO, H_2O-ice as well as PAHs (Malfait, K., Waelkens. C., Waters, L.B.F.M., Vandenbussche, B., Huygen, E. and de Graauw, M.S. 1998, Astron. Astrophys., **332**, L25).

to features at these wavelengths. There is a good agreement between the observed features in HD97048 and Elias 1 with the expected features from hydrogenated nanodiamond particles, as can be seen from Fig. 6.6. The environment around HD97048 is of high temperature and therefore contain strong UV fluxes. Hence the grains are exposed to strong UV fluxes which can anneale the preexisting diamond grains forming hydrogen monolayer similar to the laboratory studies. Therefore nano-diamond particles is a good possible carrier of the observed features.

The observed features also appear to match with the laboratory data for formaldehyde (H_2CO). However it is highly unlikely that H_2CO could

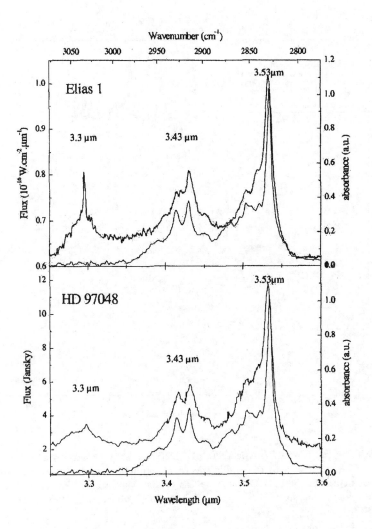

Fig. 6.6 Comparison between the observed emission spectra of Herbig Ae/Be stars, Elias 1 and HD97048 (upper curves) with the laboratory absorption spectra of nanodiamond crystals (lower curves) measured at 300K (Guillois, O., Ledoux, G. and Reynaud, C. 1999, Astrophys. J. Lett., **521**, L133: reproduced by permission of the AAS).

be present in regions where the temperature is high.

The infrared spectra in the region 3-45μm of Herbig Ae/Be stars AB Aur (AB Aurigae) and HD 163296, obtained with ISO, appears to show a feature around 23μm. This feature could be extracted from the observed spectra by subtracting the expected thermal emission from known and identified grains from the radiative transfer model. The dust components that have been considered in the model include glassy silicates, forsterite, carbonaceous material, metallic iron and water ice. After subtracting the best fit to the observed spectra the residuals are shown in Fig. 5.4. The figure clearly shows the presence of a strong and broad feature around 23μm. This is attributed to FeS. Laboratory studies show that FeS has also weaker features at 34, 38 and 44μm. These features are quite weak in the spectra but may be present.

The ISO spectrum of HD100546 shows remarkable resemblence to that of Comet Hale-Bopp indicating the similarity of the dust grains (Fig. 4.12). However there is no strong evidence for the presence of crystalline silicates in the interstellar medium. This indicates that the grains in comet must have been formed from amorphous grains during the star and planet formation. Annealing of amorphous grains require heating to temperatures above 1050K in the case of olivines. This implies that the crystalline dust in HD100546 which has now a temperature of the order of 200K or less, must have been hotter in the past which must have led to mixing of grains. It is likely that a similar process must have taken place in the proto-solar cloud resulting in the crystallisation of grains in the Comet Hale-Bopp. Exposure of dust particles to high energy cosmic rays for long periods of time can also modify amorphous to crystalline nature. Crystalline silicate and H_2O ice is common to both of these types of objects.

6.5.5.2 Vega-Type Stars

The observations carried out on several nearby stars with IRAS in the wavelength band 25 to 100μm showed the presence of excess infrared radiation above the expected photospheric emission. This gave evidence for the existence of optically thin dust shells around main-sequence stars. The star α Lyr (Vega) was the first star to show such an infrared excess radiation and hence the stars of this type are called as Vega-excess stars or Vega-Type stars. Most of the information about these objects has come from the extensive study of the prototypes, α Lyrae, α PsA (Piscis Austrini), β Pic (pictoris) and ϵ Eridani. These are spatially resolved infrared sources and show disk shape. The gas is heavily depleted in these objects and the corresponding gas to dust ratio is much smaller than unity. These have some similarity to the solar system and hence it could be 'planetary debris disks'. The grain temperature of the prototype stars is around 50 to 125K. The total mass of the observed small grains (sizes of 1-100μm) orbiting these stars is around 10^{-2} - 10^{-1} $M_\oplus$.

The evidence that the dust in these disks contains silicate came from the detection of 9.7μm emission feature in the spectrum of β Pic. The peak observed at 11.2 μm in β Pic is attributed to crystalline olivine. There is a great similarity in the silicate band profiles between Vega-Like stars and the dust in comets. The set of unidentified infrared emission features (AUIBs), have also been seen from Vega-Like stars. This shows the presence of both oxygen and carbon-rich dust species. The mm/sub mm observations of Vega-Like stars indicate that they posses grains that are much larger than those found in the interstellar medium. This could imply the growth of grains in the past or going on currently with in the discs. The Vega-Like star phenomena could be compared with the solar system's Kuiper belt in the sense that there could exist a massive cloud of comets surrounding these stars.

6.5.6 *Planetary Nebulae*

Planetary nebulae are a class of objects which are spectacular to see in the photographs. They have all very well defined spherical symmetry and in some cases with a ring structure. There is a star at its center whose temperature is quite high $\sim$ 50,000 to 150,000K. This corresponds to O-type or Wolf Rayet type of stars. The material in the shell is ionized because of the high temperature of the central star. The spectra of planetary nebulae is similar to HII regions. They show strong hydrogen and helium recombination lines. They have also strong forbidden lines arising out of OIII, NII and so on. The mechanism of excitation of forbidden lines in planetary nebulae are due to collisions with electrons. Therefore the electron temperature and electron density can be determined from the photoionized models. The typical values for electron temperature and electron densities are

$$T_e \sim 10,000 \text{ to } 15,000 K$$
$$N_e \sim 10^3 \text{ to } 10^4/\text{cm}^3. \tag{6.4}$$

The total mass in the shell $\sim 0.1 M_\odot$.

The central stars of planetary nebulae have high temperature $> 5 \times 10^4$K and high luminosity $\sim 10^4 - 10^5 L_\odot$. Stellar evolution calculations show that this is the region where highly evolved stars should lie (Appendix A, B). Hence the central stars of planetary nebulae are highly evolved stars. The material in the shell is the ejected material from the central star as the star evolves from the AGB phase to the planetary nebula stage.

The infrared observation of the bright planetary nebula NGC 7027 carried out in 1967 showed for the first time excess of radiation over and above the normal emission by the ionized gas. Since then infrared radiation has been observed in most of this type of objects and hence it is a common property of all planetary nebulae. The excess radiation seen in the infrared and far-infrared regions is attributed to thermal emission from the particles in the shell.

Since the central stars are highly evolved the material of the nebula is expected to differ from the normal abundances of elements due to nuclear processing of the material inside the star. The Wolf Rayet type of central stars is low in hydrogen but overabundant in helium, carbon, nitrogen and oxygen. Therefore it is of interest to see whether this is reflected in the composition of dust existing in planetary nebulae.

The spectra of the bright planetary nebula NGC 7027 observed in the region from 2 to 14μm in the 1970's, showed the richness of the spectra in this spectral region. This can be seen from Fig. 6.7. It revealed several strong features at 3.3, 6.2, 7.7, 8.6 and 11.3μm whose identification was not clear at that time. However it is now known that these bands belong to the well known family of bands, the AUIBs. These features have also been seen in the spectra of several planetary nebulae with Wolf Rayet type of central stars (WC type).

The ISO spectra of the two well known planetary nebulae with Wolf Rayet WC central stars, namely BD+30 3639 and He 2-113, also show strong AUIB features. In addition, features at 19.8, 23.5, 27.5 and 33.8μm are present which is due to crystalline silicate. A feature is present at 40.5μm and is attributed to crystalline pyroxene. The ISO spectra of the planetary nebula, CPD-56° 8032 (CPD), which has a late Wolf Rayet type of central star, show AUIB features, crystalline silicate features and in addition the crystalline H_2O feature at 43μm. Therefore both silicates and water ice which are oxygen-rich condensates and are not expected to form in carbon-rich outflow is present in CPD. This type of combined spectra has also been seen from several other planetary nebulae. This indicates that the presence of oxygen-rich dust around these highly carbon-rich objects is the rule rather than the exception.

A strong emission feature is also present around 65μm in MWC922 and CPD and is attributed to clinopyroxene, a crystalline silicate. Another feature present at 69μm in MWC922, CPD and Red Rectangle is due to forsterite (Mg_2SiO_4), another crystalline silicate.

The broad feature present at 21μm which is a common feature in post-

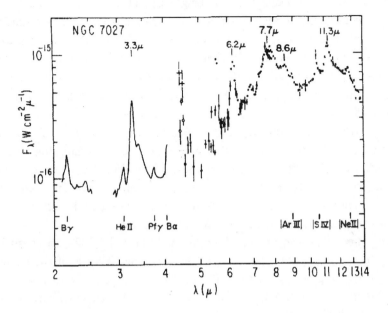

Fig. 6.7 The 2-14μm spectrum of planetary nebula, NGC7027. The strong dust emission features can be seen (Russel, R.W., Soifer, B.T. and Willner, S.P. 1977, Astrophys. J. Lett., **217**, L149: reproduced by permission of the AAS).

AGB stars (Sec. 6.4.1.) has also been seen in the planetary nebulae NGC40 and NGC6369 containing hydrogen deficient stars HD826 and HD158269. The carrier could be TiC as mentioned earlier. The detection of 21μm in planetary nebulae indicate that the carrier can survive under the extreme UV environment of the central stars of these objects. Therefore the carrier cannot be transient but has to be a stable dust component.

The feature at 30μm which is commonly seen from carbon stars are also seen from planetary nebulae. The presence of 30μm feature in the ISO spectra of a large number of planetary nebulae indicate the common nature of this feature in these objects. The radiative transfer models with the inclusion of MgS grains give reasonably a good fit to the observed feature.

The infrared spectra of the planetary nebulae M2-43 and K3-17 shows all features expected from a carbon-rich dust components such as emission from AUIBs and 30μm feature. In addition they show a feature peaking near 23μm. Laboratory measured infrared reflectance of FeS (troilite) has a broad feature at 23μm and narrower features at 33, 38 and 43μm. The laboratory spectra of 23μm feature matches well with the observed emission spectra from the two planetary nebulae. The weak features at 38 and 43μm appear to be present in the observed spectra. Therefore the carrier of 23μm feature could be FeS. FeS is also predicted to form in the outflow of carbon-rich stars.

From the above discussion it is clear that the spectra of planetary nebulae show silicate emission at longer wavelengths indicating that the oxygen-rich dust is at lower temperatures. This suggest that they are located farther from the central star compared to the carbon-rich dust. This could be explained in terms of recent transition from an oxygen-rich mass loss to carbon-rich outflow as the star evolved into carbon-rich objects.

6.5.7 Novae

Some stars brighten up for a few months or years before fading again. The characteristic feature of the light curve is that there is a very rapid brightening by eight to ten magnitudes over one or two days. The rapid rise is followed by a decline of a few magnitudes in three or four weeks and then return to the pre-outburst brightness over the next one or 10 years. The spectrum of a nova changes rapidly during an outburst. Initially the absorption lines seen in the visible region is characteristic of hot stars of spectral type B or A. The spectral type then move towards that of the cooler stars and emission lines start to appear. In the final stages the spectrum is dominated by emission lines. These are called classical nova. There is another class of novae called recurrent novae which have light curves similar to the classical novae but several outbursts occur. There is often evidence that the star is actually a close binary system.

Nova outburst arises as a result of mass accreting from the companion star on to the white dwarf. This leads to very high temperatures which trigger thermonuclear ignition. As a result the matter is ejected out as an expanding shell of ionized gas (Appendix B).

The formation of graphite grains in the nova ejecta was considered in the 1970s. Theoretical studies indicate that grain condensation takes place in the expanding material when the temperature cools to around 1000K. It is likely that condensation takes place via a carbon-rich chemistry leading to condensations such as, carbon, graphite etc.

The infrared studies of several novae has shown that nearly all of them develop rapid rise at infrared wavelengths about 3 to 4 months after the outburst. This is accompanied by the steep decline in the UV and visual wavelengths. This behaviour of the light curve has been interpreted as evidence for condensation and growth of dust particles in the novae ejecta. A complete study of the infrared development of several novae has shown the presence of three discrete phases in their shell evolution. Initially the expanding shell is optically thick as the ejecta moves out. As the shell expands it becomes optically thin and free-free emission is observed. Sometime during this phase several strong features appears in emission. In the third phase dust formation takes place. Several novae have been monitored continuously for understanding the formation and nature of dust in the ejecta.

A common feature of several novae which has been observed in the spectral range 8 to 13 μm is that they show evidence for emission from at least two dust components. A featureless continuum which is generally attributed to carbonaceous material like graphite and a broad feature at 10μm which is attributed to emission from silicate material. The silicate feature usually becomes stronger with time. The observation of Nova Aquilae 1982 showed the ratio of fluxes at 9.7μm contributed by the silicate and to smooth grain components, increased from $\sim$ 2 on day 143 to $>$ 10 on day 276. It was also noted that most of the luminosity emitted by Nova Aql was in the infrared region after about day 35. Extreme abundance anomalies such as neon is over-abundant by $\sim$ 730, CNO overabundant by factors of about 24, 32 and 22 respectively, are seen.

The infrared photometry of Nova Vulpeculae 1984 (NV2) from 2.3 to 19.5μm showed the 19μm silicate emission feature along with the normal 9.7μm seen in other novae. The observation showed that the silicate emission features increased in intensity by at least a factor of 2 between May 15 and August 23 due to grain condensation and growth. The analysis of 12.8μm [NeII] emission line from NV2 also showed over abundance of Ne in the ejecta.

The mid-infrared spectrum (8 to 13μm) of Nova Cen 1986 showed usual presence of silicate and carbon dust particles. In addition most of the AUIB features, namely 3.28, 3.5, 6.2, 7.7, 8.6 and 11.3μm were detected. The presence of both carbon and silicate grains in the ejecta means that there must be substantial chemical gradients within the ejecta. Abundance gradients that cause the C/O ratio to change from $>$ 1 to $<$ 1 could produce layers or shells of carbon and silicate type of dust particles. The estimated variation in grain temperature for dust particles of carbon and silicate varied between 600 to 300K and 400 to 100K respectively between day 100 and day 900 in Nova Cen 1986. The corresponding dust mass of graphite and silicate is around 5×10^{-8} $M_\odot$ and 2×10^{-7} $M_\odot$ respectively. This dust mass is smaller than that for Nova Aql 1982 by more than a factor of 10 or so.

The enhanced abundance of different elements seen in different novae indicate the possible presence of carbon-oxygen (C-O), oxygen-neon-magnesium (O-Ne-Mg) and other types of white dwarfs accreting matter in binary systems. This could result in various kinds of composition for the silicate and carbon type of dust particles.

6.5.8 Supernovae

The two types of Supernovae, Type I and Type II are classified according to their light curve. Type I Supernova (SNeIa) are more spectacular reaching a peak absolute magnitude near -19. After a rise in brightness it declines at an initial rate of one magnitude per week and later slows down to one magnitude in 10 weeks or so. The spectra is characterised by the absence of lines due to hydrogen. This occurs in all types of galaxies. Supernova Type Ia is believed to arise from the thermonuclear explosion of the core of the accreting white dwarf after passing through AGB and planetary nebula stage (Appendix B).

In Type II Supernovae (SNeII) the peak brightness is lower with typically absolute magnitudes $\sim$ -12 to -16. They tend to remain at their peak brightness for longer than Type I Supernovae and then decline more slowly. The spectra contain strong hydrogen lines. They are observed to occur exclusively in the spiral arms of spiral galaxies. Supernova Type II arises from massive stars (> 8 $M_\odot$) which cannot give rise to AGB stars (Appendix B).

The idea that dust particles could be formed in supernovae was proposed in 1970s. The type of condensation depends crucially on the assumption made regarding the composition and type of mixing in the supernova ejecta. Since ejecta consists of a number of regions of different chemical composition and each of these regions would therefore lead to condensation of different grain type. Therefore in the theoretical calculation the star leading to supernova explosion is assumed to have a 'Onion-like' structure. i.e. the layers arising out of successive nuclear burning stages of the central star. Five layers referred to as (H), (He), (C), (O) and (Si) i.e burning zones were considered. Chemical equilibrium calculations were carried out for each of the layers separately for appropriate temperature, pressure, abundance ratio of elements etc. The resulting composition of condensates depend on the nature of each layer. For example in carbon rich region it leads to carbon and graphite grains. In a metal rich region it leads to the condensation of silicates such as $MgSiO_4$ or Mg_2SiO_4 or to Al_2O_3 or to magnetite, Fe_2O_3 or to metallic Fe-Ni grains etc. But overall on the average they are typically oxygen-rich condensates.

The possibility of condensation of carbon grains in the O-rich interior of an expanding Supernova Type II has been investigated. In the O-rich interior, all C is tied up as CO. Therefore if carbon grains have to form somehow C has to be freed from CO. This has been brought about by an elegant mechanism.

After the explosive event, the interior contain highly energetic particles. This is achieved from the radiation eminating from the ^{56}Co radioactivity. The atoms C and O are then produced from the dissociation of CO molecule by energetic electrons. Carbon now exists in free form. This makes the carbon vapour pressure very high compared to the thermodynamic equilibrium value corresponding to the kinetic temperature of the gas. The presence of supersaturation vapour pressure of carbon can lead to condensation of carbon into solid form. For this process to take place the free carbon has to first form linear chain, C_n molecules. When n becomes sufficiently large ($n \sim 10\text{-}14$), the linear chain spontaneously close to form a ring. These then act as nucleation centers for the growth of macroscopic grains. Chemical network for the growth of carbon chain taking into account formation and destructive mechanisms has been performed for number densities of C and O of $10^{10}/cm^3$. This yields the steady state abundances of each linear chain carbon molecules. This has been followed by the growth process. It is shown that the proposed mechanism can lead to large graphite grains of sizes seen as presolar graphite grains in meteorites.

These studies clearly show the sensitivity of the resulting dust composition to model parameters and assumptions involved. Therefore the type of dust particles that could form in a supernova ejecta should depend on the nature, structure and evolutionary status of the star.

Type Ia Supernova produce mostly Fe. The amount is much more than it can bind with O or S to form oxides and sulphides. Therefore much of the iron from SNeIa may condense as pure iron. This is in contrast to Type II supernovae where more O and Mg are produced relative to Fe (i.e. O/Fe>1).

The direct way to look for dust in SNeII is to look for thermal emission from dust particles in the infrared region. The first infrared measurements carried out in the near infrared region of SN 1979c and SN 1980k showed mixed results. However, SN 1987A in the LMC provided the opportunity for the study of dust ejecta. The continuous monitoring of SN 1987A in the infrared wavelength region provided the direct evidence for the formation of dust in supernova ejecta.

The formation of dust in 1987A occured around day 530 after the explosion came out of several considerations. When the infrared excess in the near and mid-infrared regions was observed there was a corresponding drop in the UV and visual fluxes. The emission line profiles of various elements in the ejecta which was seen as expected from a spherically expanding shell developed asymmetry with the onset of dust formation. The intensities of various emission lines in the visual region such as [MgI] and [OI] decreased in intensity and could be attributed to extinction by the newly formed dust particles in the ejecta. There was also depletion of various refractory elements from the gas phase. The estimated optical extinction ~ 0.7 mag. In the mid-infrared spectral region features usually attributed to silicates, AUIBs, amorphous carbon or SiC were not present. Since the emission is optically thick around $10\mu m$, the absence of silicate feature around $10\mu m$ does not necessarly exclude silicate condensation (Fig. 4.14). Such a condensation may be implied by the detection of SiO band $\triangle V = 1$ at $7.8\mu m$. Since the dust emission is best fitted by a grey body it is rather difficult to infer the composition of the dust.

In summary, dust condenses in diverse conditions starting from cool stars to expanding supernovae. Therefore dust is associated with every object in the universe. Hence dust is universal. The dust produced is mainly of carbonaceous and silicate in nature. Silicates exhibit a wide range in their mineralogy. It is remarkable that some of the products such as dust from circumstellar shells of stars and novae and the explosive nucleosynthesis products of supernovae are preserved in meteorites and interplanetary dust particles even though they represent two different episodes separated by vast interval of time.

References

Some review articles are the following
Habing, H.J. 1996, Astron. Astrophys. Rev., **7**, 97.
Willson, L.A. 2000, Ann. Rev. Astron. Astrophys., **38**, 573.
Winckel, H.V. 2003, Ann. Rev. Astron. Astrophys., **41**, 391.
Zuckerman, B. 2001, Ann. Rev. Astron. Astrophys., **39**, 549.

Methods of detecting mass loss

Knapp, G.R. 1991, *Frontiers of Stellar Evolution*, ASP Conference Series, No.20, p.229.

Larmers, H.J.G.L.M. and Cassinelli, J.P. 1999, *Introduction to Stellar Winds*, Cambridge University Press, Cambridge.

Weymann, R. 1963, Ann. Rev. Astron. Astrophys., **1**, 97.

Particle Condensation

Cherchneff, I., Le Teuff, Y.H., Williams, P.M. and Tielens, A.G.G.M. 2000, Astron. Astrophys., **357**, 572.

Donn. B., Wickramasinghe, N.C., Hudson, J.P. and Stecher, T.P. 1968, Astrophys. J., **153**, 451.

Ferarotti, A.S. and Gail, H.-P. 2002, Astron. Astrophys., **382**, 256.

Gail, H.-P. and Sedlmayr, E. 1999, Astron. Astrophys., **347**, 594.

Gilman, R.C. 1969, Astrophys. J., **155**, L185.

Hoyle, F. and Wickramasinghe, N.C. 1962, Mon. Not. Roy. Astron. Soc., **124**, 417.

Kamijo, F. 1963, Publ. Astron. Soc. Japan, **15**, 440.

Lodders, K. and Fegley Jr., B. 1999, In *Asymptotic Giant Branch Stars*, IAU Symposium No.191, Eds. T.L. Bertre, A. Lebre and C. Waelkens, ASP Publishers, p.279.

Tsuji, T. 1973, Astron. Astrophys., **23**, 411.

IRC+10216

Cernicharo, J. 2000, In *Astrochemistry: From Molecular Clouds to Planetary System*, IAU Symposium No.197, Eds. Y.C. Minh and E.F. van Dishoeck, ASP Publishers, p.375.

Glassgold, A.E. 1999, In *Asymptotic Giant Branch Stars*, IAU Symposium No.191, Eds. T.Le. Bertre, A. Lebre and C. Waelkens, ASP Publishers, p.337.

Observational Results.

SiC

Hackwell, J.A. 1972, Astron. Astrophys., **21**, 239.

Little-Marenin, I.R. 1986, Astrophys. J., **307**, L15.

Speck, A.K., Barlow, M.J. and Skinner, C.J. 1997, Mon. Not. Roy. Astron. Soc., **288**, 431.

Sylvester, R.J. 1999, In *Solid Interstellar Matter: The ISO Revolution*, Eds. d'Hendecourt, L. Joblin and C. Jones, Springer-Verlag, p.263.

TiC

Helden, G.V., Tielens, A.G.G.M., van Heijnsbergen, Duncan, M.A., Hony, S., Waters, L.B.F.M. and Meijer, G. 2000, Science, **288**, 313.

Li, A. 2003, Astrophys. J., **599**, L45.

Silicates
Low, F.J. and Krishna Swamy, K.S. 1970, Nature, **227**, 1333.
Molster, F.J., Waters, L.B.F.W., Tielens, A.G.G.M. and Barlow, M.J. 2002, Astron. Astrophys., **382**, p.184, p.222 and p.241.
Sylvester, R.J., Kemper, F., Barlow, M.J., de Jong, T, Waters, L.B.F.M., Tielens, A.G.G.M. and Omont, A. 1999, Astron. Astrophys., **352**, 587.

FeSi
Damascelli, A., Schullte, K. and van der Marel, D. 1997, Phy. Rev., **B55**, R4863.
Ferrarotti, A., Gail, H.P., Degiorgi, L. and Ott, H.R. 2000, Astron. Astrophys., **357**, L13

Herbig Ae/Be star HD100546
Bouwman, J., de Koter, A., Dominik, C. and Waters, L.B.F.M. 2003, Astron. Astrophys., **401**, 577.
Malfait, K., Waelkens, C, Waters, L.B.F.M., Vandenbussehe, B., Huygen, E. and de Graauw, M.S. 1998, Astron. Astrophys., **332**, L25.

Vega-like stars
Aumann, H.H. 1985, Pub. Astron. Soc. Pacific, **97**, 885.
Aumann, H.H., Gillett, F.C., Beichmann, A., Gillett, F.C., de Jong, T., Houck, J.R., Low, F.J., Neugebauer, G., Walker, R.G. and Wesselius, P.R. 1984, Astrophys. J., **278**, L23.
Beckman, D.E. and Paresce, F. 1993, In *Protostars and Planets III*, Eds. E.H. Levy and J.I. Lunine, University of Arizona Press, Tucson, p.1253.
Coulson, I.M. and Walther, D.M. 1995, Mon. Not. Roy. Astron. Soc., **274**, 977.
Sylvester, R.J., Skinner, C.J., Barlow, M.J. and Mannings, V. 1996, Mon. Not. Roy. Astron. Soc., **279**, 915.

Planetary Nebulae.
First IR detection
Gillett, F.C., Low, F.J. and Stein, W.A. 1967, Astrophys. J., **149**, L97.

First Dust modeling
Krishna Swamy, K.S. and O'Dell, C.R. 1968, Astrophys. J., **151**, L61.

Other work
Dijkstra, C., Waters, L.B.F.M., Kemper, F., Min, M., Matsuura, M., Zijlstra, A., de Koter, A. and Dominik, C. 2003, Astron. Astrophys., **399**, 1037.
Habing, H.J. and Lamers, H.J.G.L.M. 1997, IAU Symposium No.180, Kluwer Academic Publishers, Dordrecht.
Kemper, F., Molster, F.J., Jager, C. and Waters, L.B.F.M. 2002, Astron. Astrophys., **394**, 679.
Kwok, S., Volk, K.M. and Hrivnak, B.J. 1989, Astrophys. J., **345**, L51.

Waters, L.B.F.M., Beintema, D.A., Zijlstra, A.A., de Koter, A., Molster, F.J., Bouwman, J., de Jong, T., Pottasch, S.R. and de Graauw, Th. 1998, Astron. Astrophys., **331**, L61.

Novae.
First to consider carbon condensation
Clayton, D.D and Hoyle, F. 1976, Astrophys. J., **203**, 490.

Later work
Gehrz, R.D., Grasdalen, G.L., Greenhouse, M., Hackwell, J.A., Hayward. T and Bentley, A.F. 1986, Astrophys. J., **308**, L63.
Roche, P.F., Aitken, D.E. and Whitmore, B. 1984, Mon. Not. Roy. Astron. Soc., **211**, 535.
Shore, S.N. and Gehrz, R.D. 2004, Astron. Astrophys., **417**, 695.
Smith, C.H., Aitken, D.K. and Roche, P.F. 1994, Mon. Not. Roy. Astron. Soc., **267**, 225.

Supernovae.
First to consider particle condensation
Lattimer, J.M., Schram, D. and Grossman, L. 1978, Astrophys. J., **219**, 230.

Other papers
Cernuchi, F., Marsieano, F and Codina, S. 1967, Ann d'Ap., **30**, 1039.
Clayton, D.D., Deneault, E.A.N. and Meyer, B.S. 2001, Astrophys. J, **562**, 480.
Fischera, Tg, Tufts, R.J. and Volk, H.J. 2002, Astron. Astrophys., **395**, 189.
Hoyle, F. and Wickramasinghe, N.C. 1970, Nature, **226**, 62.
Kozasa, T., Hasegama, H. and Normoto, K. 1991, Astron. Astrophys., **249**, 474.
Wooden, D.H., Rank, D.M., Bregman, J.D., Witteborn, F.C., Tielens, A.G.G.M., Cohen, M., Pinto, P.A. and Axelrod, T.S. 1993, Astrophys. J. Suppl., **88**, 477.

MgS
Hony, S., Waters, L.B.F.M. and Tielens, A.G.G.M. 2002, Astron. Astrophys., **390**, 533.

FeS
Hony, S., Bouwman, J., Keller, L.P. and Waters, L.B.F.M. 2002, Astron. Astrophys., **393**, L103.

Chapter 7

Extragalactic Dust

7.1 Introduction

Galaxies come in various types and shapes. They can be classified broadly into two types, normal and irregular types. Normal galaxies contain both ellipticals and spirals. Hubble classified these normal galaxies in a sequence based on their structural appearence. It starts with Elliptical galaxies containing perfectly spherical ones (E0) and ending with the highly flattened ones (E7). At the end of the flattened system, it bifurcates into two divisions labelled as spirals and barred spiral galaxies. Spiral galaxies are classified in order of tightness of the winding of their arms from the most tightly wound (type Sa) to the most loosely wound (Sc). Barred spirals are ordinary spirals except that they have a luminous bar running through their nucleus (SBa to SBc).

The irregular galaxies are galaxies that do not readily fit into the Hubble sequence. They come in most unusual shapes. The distorted shape could arise due to tidal forces of the nearby larger galaxy or colliding with another galaxy or due to some gigantic explosion taking place inside.

All the above galaxies have very different internal motions, types of stars, metallicities, light distributions and so on. The reason for their observed differences may partly be due to different star formation histories. In particular ellipticals formed most of the stars early on leaving less than a percent of their mass in the form of gas today, while spirals still have around 5 to 15 percent of their mass in gas which can form new stars.

In the later part of the century, with the availability of new instruments and techniques, a large number of other types of galaxies have been discovered which donot fall into Hubble sequence. These are called Active galaxies and characterised by unusually energetic activities. They include

Radio galaxies, Seyfert galaxies and Quasars. Radio galaxies emit much of their energy in the radio region. The Seyfert galaxies are intrinsically much brighter than ordinary galaxies. Their energy output exceeds that of giant elliptical galaxies by atleast a factor of 10 to 100. The energy is emitted not only in the radio part of the spectrum but also in infrared, optical, UV and X-ray region. They are all often powerful radio sources. Most of this energy originates from a very compact region at its centre. Quasars or quasi-stellar objects are another group of objects which are characterised by large redshifts of the lines in their visible spectra. Redshift is generally denoted by the dimensionless quantity z. Since the observed wavelength of the light from distant galaxies is shifted to longer wavelength than the wavelength at which the light had been emitted, z represents the ratio between this wavelength shift to the wavelength at which the light was emitted. Quasars appear as point- like sources but abnormally luminous. They emit intense radiation from X-rays to gamma-rays.

The broadband infrared observations carried out in 1970's and 1980's had indicated that all galaxies emit much more in the infrared compared to visible wavelength region. Infrared observations carried out with Kuiper Airborne Observatory (KAO) and the launching of several satellites such as, Infrared Astronomical Satellite (IRAS) and Infrared Space Observatory (ISO) devoted to studies in the infrared region, has increased the knowledge about dust in external galaxies in a dramatic way. The Infrared Space Observatory with high spectral resolution and covering the full wavelength region 2-200μm, revealed the presence of several dust spectral features which will be discussed in individual cases. The study has also made great strides due to HST, Keck, SCUBA, IUE, COBE and X-ray satellites, EXOSAT, Ginga, Einstein, ASCA, etc. So the entire electromagnetic spectrum from X-rays to radio region is used to investigate the nature of dust in and around extragalactic systems. Combining these with highly sensitive instruments and ingenuity in observational techniques has led to phenomenal progress in the understanding of dust in the universe. All these have made it possible to study the nature of dust at redshifts in excess of 6 when the universe was hardly 6% of its current age.

There is lot of evidence for the presence of dust in almost all the external galaxies. The striking evidence for the dust in external galaxies particularly in spirals and irregulars comes from their photographs. They show clearly the dust lanes and patches in a large fraction of galaxies. The opaque dust layer present in the galaxy NGC 4594 can clearly be seen in the photograph shown in Fig. 7.1. Similarly, Fig. 7.2 show the photograph of the spiral

Fig. 7.1 Hat Galaxy NGC 4594 is a spiral galaxy in Virgo, seen edge on. The prominent dust lane through central plane can clearly be seen.

galaxy M51 seen face on. The integrated light from such galaxies show linear polarization indicating the presence of alligned grains.

Here we would like to discuss briefly the studies relating to dust in extragalactic systems by taking some representative cases.

7.2 Magellanic Cloud

The two nearest irregular galaxies to the milky way galaxy are the Large Magellanic cloud (LMC, ~52Kpc) and Small Magellanic cloud (SMC, ~63Kpc). These two galaxies are members of the Local Group. Because they are close by it is possible to make various kinds of observations on these two galaxies. Both these galaxies have been studied based on individual stars, HII regions and in the radio region. One of the main aim of these studies is to see whether the dust properties in these two galaxies are similar to the Milky Way. Abundances of elements have been determined from the study of emission lines in HII regions. They seem to indicate that the abundance of metals (including C, N, O) relative to hydrogen are considerably less than Galactic values. Therefore these elements may be locked up in grains.

Fig. 7.2 M51, a spiral galaxy seen face-on. The dust lanes in spiral arms can be seen.

7.2.1 Extinction Curve

The usual method of getting the extinction curve is to compare pairs of reddened and unreddened stars of similar spectral type. Unfortunately, the scarcity of known reddened stars and their apparent weakness is the main difficulty with extinction studies in the SMC. The situation with regard to LMC is better as a larger number of reddened stars have been used. The early observations in the U, B, V range and infrared observations upto around 2μm had shown that the interstellar extinction curve in both SMC and LMC were essentially the same and was also very similar to the interstellar law in the Galaxy. Therefore there is no significant difference between the LMC, SMC and the Galaxy as far as the visible and infrared extinction is concerned. This may imply the similar population of relatively larger size grains.

The extinction curve was extended into the UV region based on the observations carried out with the IUE satellite in the low resolution mode. The resulting extinction curve for these two galaxies are shown in Fig. 7.3. The extinction curve for the Galaxy is also shown in the figure for comparison, which differ from those of LMC and SMC in the UV region. LMC has a weaker 2175Å feature and a weaker far-ultraviolet extinction rise. In the SMC, the feature at 2175Å is absent and has a sharper rise in the far-ultraviolet extinction. Therefore there is a trend of increase in strength of the 2175Å feature going from SMC to LMC to Galaxy. This result indicates that the carriers responsible for 2175Å feature is less in LMC compared to the Galaxy and more so in SMC. This is consistent with the underabundance of carbon by a factor of 30 or so in SMC compared to that of Galaxy. The observed variation in the average metallicities in SMC, LMC and Galaxy of around 0.2, 0.4 and 1.0 in solar units is similar to the trend seen in Fig. 7.3. The observed trend suggest that the dust composition and abundance may depend upon galactic star formation history and stellar initial mass function. This will have effect on the observed reddening in galaxies. The value of R_v (A_v/E_{B-V}) varies from 2 to 7 among different galaxies.

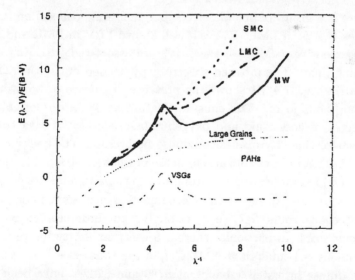

Fig. 7.3 Shows the average extinction curves for Milky Way (MW), Large and small Magellanic Clouds (LMC and SMC). The contribution from three dust components are also shown (Sauvage, M. 1997, In *Interstellar Medium in Galaxies*, Ed. J.M. van der Hulst, Kulwer Academic Publishers, p.1: with kind permission of Springer Science and Business Media).

The polarization studies in the wavelength region 0.35 to 0.84μm have been carried out on a number of reddened stars in LMC. The dependence of observed polarization in LMC appears to be similar to that found in the Galaxy. The wavelength of maximum polarization, λ_{max} ~0.55μm is similar to that of Galaxy. Therefore the results based on the studies of wavelength dependence of polarization of a large sample of stars in LMC indicate that the properties of dust grains must be quite similar to the dust grains found in the Galaxy.

The polarization measurements have also been carried out on a limited sample of stars in SMC. The observed wavelength of maximum polarization, λ_{max}~0.45±0.08μm. This is smaller than the Galactic average of λ_{max}~0.55μm. This is in contrast to the results of LMC. The smaller value of λ_{max} in SMC indicate the presence of smaller average grain sizes in SMC compared to that in the Galaxy.

7.2.2 Spectral Features

Earlier observations in the spectral region 8-12μm of the HII region N44A in LMC and N88A in SMC had shown the presence of emission feature only in LMC indicative of silicate feature.

There is a group of hot massive supergiant stars called Luminous Blue Variables (LBV) in LMC. R71 is a well known LBV in LMC and has been studied extensively. Most of the LBVs are associated with ring structure containing dust. The infrared spectroscopy carried out with ISO on R71 is shown in Fig. 7.4. It shows the presence of silicate in both crystalline and amorphous form. The emission feature at 23.5μm is often found in the spectra of cool stars and oxygen-rich envelopes. It is attributed to crystalline olivine. Pyroxenes could also be present. The emission features at 6.2, 7.7, 11.3 and possibly at 8.6μm are present (AUIBs). The presence of AUIBs in R71 is interesting as carbon bearing small grains are not expected to be formed in an oxygen-rich environment. The observed composition of the dust shell around R71 is quite similar to those seen in galactic red supergiant stars. In particular there is a great similarity between R71 and the luminous red supergiant NML Cyg in the Galaxy.

The diffuse interstellar bands at 5780 and 5797Å have been detected towards reddened stars of SMC. The line at 6284Å towards several reddened stars in the LMC show variation in their strength. Therefore the carrier responsible for diffuse interstellar bands may depend on the environment properties.

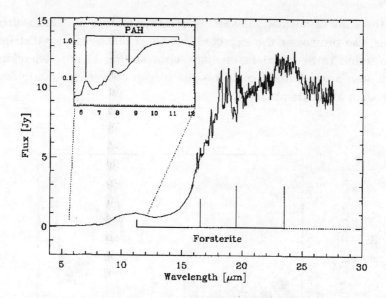

Fig. 7.4 The spectrum of Luminous Blue Variable R71 in the Large Magellanic cloud. The insert shows the 6 to 12μm region with the AUIB emission features. The wavelength of the expected bands of forsterite (Mg_2SiO_4), a crystalline silicate is also shown (Voors, R.H.M., Waters, L.B.F.M., Morris, P.W. et al. 1999, Astron. Astrophys., **341**, L67).

The observations carried out with the camera ISOCAM on board ISO in the wavelength region 5-16μm of a molecular cloud in SMC has shown the presence of emission features at 6.2, 7.7, 8.6, 11.3 and 12.7μm (AUIBs). The features at 6.2, 7.7 and 11.3μm are much stronger than the features at 8.6 and 12.7μm. These observations indicate that the carrier responsible for these features are present in abundance in SMC.

The gas-to-dust ratio ($N_H/E(B-V)$) in SMC $\sim 4.5 \times 10^{22}$ atom/cm^2/mag, about 10 times the Galactic value. In LMC the ratio is around 2×10^{22} atom/cm^2/mag about 4 times the Galactic value.

7.3 Normal Galaxies

A large number of normal galaxies have been observed with IRAS and ISO. The mid- infrared spectra in the region 5-20μm of all these galaxies are dominated by emission features at 6.2, 7.7, 8.6 and 11.3μm which are commonly seen from the interstellar medium and are generally attributed to aromatics. The average spectra determined from the observation of several galaxies is shown in Fig. 7.5. This should represent the emission features expected from the ISM of normal galaxies. The other feature at 3.3μm although weak is also present at the expected wavelength. The energy distribution is also found to be invarient in shape upto around 11μm. Therefore the trough seen around 10μm could be due to gap between aromatic features rather than a silicate absorption.

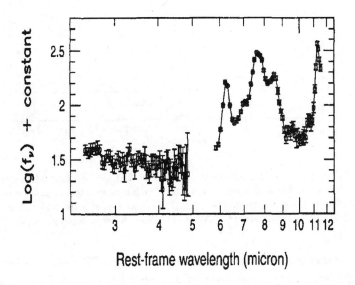

Fig. 7.5 Shows the average spectrum in the wavelength region 2-12μm derived from the study of 28 galaxies (Helou, G., Lu, N.Y., Werner, M.W., Malhotra, S. and Silbermann, N. 2000, Astrophys. J., **532**, L21: reproduced by permission of the AAS).

The mid-infrared wavelength region has also been used extensively as a diagnostic tool for the identification of the nature of galaxies. This comes from the study of the so called ISO-IRAS colour-colour diagram. In this diagram, the flux ratio F (6.75μm)/F (15μm) from ISO is compared with IRAS flux ratio F (60μm)/F (100μm) as shown in Fig. 7.6. The flux in the 6.75μm band is dominated by AUIB feature, while larger size dust particles are responsible for the observed emission in the far-infrared region. For large range of F (60μm)/F (100μm) colours, the F (6.75μm)/F (15μm) colour is roughly constant. This is the region occupied by normal galaxies. It is only beyond the IRAS colour ≥-0.3 that the ratio F (6.75μm)/F (15μm) colour decreases. This is the region where blue compact interacting or starburst galaxies occupy the diagram. This decrease arises due to the high radiation field in these objects which destroy the carrier of AUIBs. Even in the energetic regions of the centre of the Galaxy, HII region etc, where the radiation field is of orders of magnitude larger than the local interstellar field, the strength of AUIB emission features decreases drastically arising due to destruction of the carrier.

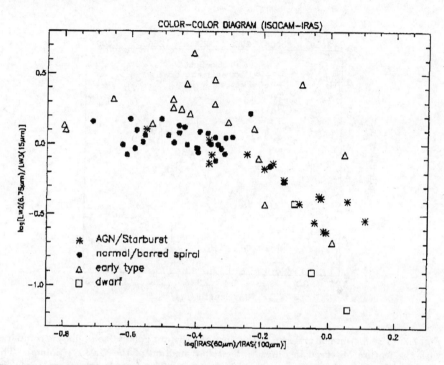

Fig. 7.6 IRAS-ISO colour diagram for galaxies. The decrease in the ratio F $(6.75\mu m)/F(15\mu m)$ arise due to the destruction of AUIB carriers in the intense radiation field and due to the increased emission from very small grains (Vigroux, L., Charmandaris, V., Gallais, et al. 1999, In *The Universe as Seen by ISO*, ESO SP-47, ESA Publication Division, ESTEC, Noordwijk, The Netherlands, p.805).

The non-stellar emission in the near-infrared region (1-5μm) has an average colour temperature $\sim 10^3$K. This is generally attributed to emission from very small grains. This infrared flux is roughly a few percent of the far-infrared flux in the region 40-120μm which arise due to thermal emission by large size grains. The near-infrared flux is found to be roughly proportional to the strength of the AUIBs. This indicates that the observed near-infrared flux originate in the interstellar medium of galaxies. It also shows that the near-infrared flux and AUIBs arise from similar carrier. The total luminosity of AUIBs in the region 5.8 to 11.3μm is around 10-20%

of the far-infrared emission. The relative strength of AUIBs vary on the average between 15 to 25%. This variation could be attributed to factors such as radiation field etc.

The spectrum of the 3.4μm feature for the dust from the extragalactic luminous IRAS galaxy 08572+3915 is shown in Fig. 7.7. For comparison

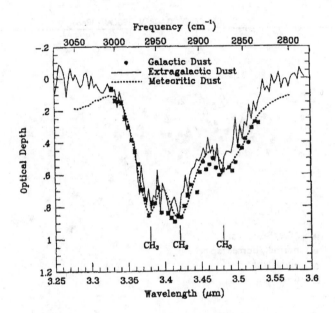

Fig. 7.7 The 3.4μm feature seen in the dust of a distant galaxy (solid line) is compared with the feature observed in diffuse interstellar medium of the Galaxy (points). The similarity between the curves can clearly be seen. The feature from the Murchison meteorite is also shown (Pendleton, Y.J. 1996, In *The Cosmic Dust Connection*, Ed. J.M. Greenberg, Kluwer Academic Publishers, p.71: with kind permission of Springer Science and Business Media).

the same feature observed in our Galaxy is also shown. The extragalactic dust feature has been corrected for the Doppler shift due to the expansion velocity of the galaxy. The close match in position, strength and profile between the galactic and extragalactic dust features suggest that this type of chemical arrangement may be quite common.

The far-infrared observations of a number of normal galaxies have been carried out with IRAS and ISO. The ISO observations extending upto 200μm has shown the presence of cold dust component with typical temperature $\sim$20K (10-28K). This is about 10K lower than the temperature derived from IRAS observations. This is due to the fact that IRAS wavelength coverage is upto 100μm and the contribution from cold dust is appreciable much beyond this wavelength. The inferred gas-to-dust ratio of this cold dust component is similar to the Galactic value.

7.4 Seyfert Galaxies

Seyfert galaxies are a distinct class of galaxies which have very bright nucleus. These nuclei possess strong and broad permitted and forbidden lines of elements. These galaxies are broadly classified into two classes, Seyfert 1 and Seyfert 2. In Seyfert 1 permitted and forbidden lines originate from distinctly different regions, while in Seyfert 2 they are formed in the same region of the nucleus. The observed differences between Seyfert 1 and 2 is believed to arise due to different orientations of the same object to the line of sight.

A large number of Seyfert galaxies of type 1 and 2 have been observed in the infrared region. The average spectra of Seyfert 1 and Seyfert 2 galaxies in the wavelength region 5.5 to 11.5μm is shown in Fig. 7.8. Figure 7.8 (bottom) clearly show the presence of strong AUIBs with peaks at 6.2, 7.7 and 8.6μm in Seyfert 2 galaxies. The feature at 3.3μm is weak but has been seen in most of the sources. The 11.3μm feature could also be present which is inferred from the nature of the observed flux in this wavelength region. As can be seen from Fig. 7.8 (top), the AUIBs are also present in Seyfert 1 galaxies but they are much weaker than that of Seyfert 2 galaxies. Figure 7.8 show that the mid-infrared spectra of Seyfert 1 and Seyfert 2 are vastly different. Seyfert 1 galaxies have weak AUIB features and a strong continuum. But Seyfert 2 galaxies have strong AUIB features and a weak continuum. The observed 7.7μm feature is much stronger in Seyfert 2 than in Seyfert 1 galaxies. This difference is not due to Seyfert 1 having weaker 7.7μm emission but is due to the presence of stronger continuum compared to Seyfert 2. The distribution of AUIB luminosities and the ratio of luminosities in the AUIBs to far-infrared region are found to be similar for both the types of galaxies. The silicate emission feature at 9.7μm is present in Seyfert 1 spectrum. However for Seyfert 2 galaxies it is rather difficult to place the continuum in the 9.7μ region but the silicate feature is likely to be present.

IRAS and ISO data has shown that the far-infrared energy distribution of Seyfert galaxies are dominated by thermal dust emission. The observed spectral distribution in the wavelength region 4-200μm could be fitted with

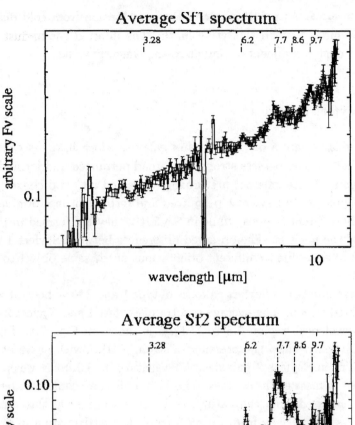

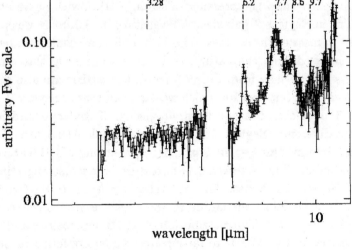

Fig. 7.8 Average ISO spectra of 20 Seyfert 1 galaxies (top) and 23 Seyfert 2 galaxies (bottom). The wavelength of the features are marked. (Clevel, J., Schulz, B., Altieri, B. et al. 2000, Astron. Astrophys., **357**, 839).

thermal emission from three different sources. The grain temperature corresponding to these three components are, ∼150K (warm component), 40-50K (cold component) and 10-20K (very cold component). The grain temperature corresponding to warm component tends to be lower in Seyfert 2 galaxies compared to that of Seyfert 1 galaxies. These three components seems to indicate well separated spatial regions. The three temperature ranges (warm to very cold) is attributed to arise from dust close to the nucleus, dust heated in star forming regions and dust heated by the general galactic interstellar radiation field.

7.5 Starburst Galaxies

Starburst galaxies constitute an important class of objects. In such galaxies the star formation in the central regions is very active and takes place in the form of short-lived 'starburst'. The evidence for this comes from the presence of hot-spots in the central region of these galaxies. These galaxies are copious emitters of infrared radiation arising from dust heated by radiation from hot young stars and severally obscured at UV and visible wavelengths. M82 (3Mpc) is considered to be a prototype of this class of galaxies and is one of the best studied starburst galaxies.

The ground based infrared observations and the observations carried out with Kuiper Airborne Observatory of M82 in 1970's had already shown the presence of well known emission features at 3.3, 6.2, 7.6, 8.7 and 11.3μm. This can be seen from Fig. 7.9. These observations have been confirmed based on the observations with ISO. Therefore M82 is characterized by the prominent AUIB features. The dip present in the 10μm region in the mid-infrared spectral energy distribution of M82 appears like a 9.7μm silicate absorption feature. However the observed spectrum of M82 can be fitted extremely well by the superposition of the spectrum of the reflection nebula NGC7023 having strong AUIB features and the continuum produced by very small grains. In essence, the dip around 10μm is caused by the strong AUIB emission on either side of the silicate feature. Therefore no silicate absorption is required. This is also consistent with the lack of clear signature of the corresponding 20μm silicate absorption feature. The broad absorption feature present around 3.0μm could be attributed to H_2O-ice. The optical depth of this feature ~ 0.2 which is smaller than the Galactic value ~ 0.5. Since the overall extinction in the two cases differ in the same sense it indicates that the line of sight properties of M82 could be similar to that towards the Galactic centre in having diffuse interstellar medium as well as molecular clouds that can host the icy grains. These properties are quite typical of starburst galaxies.

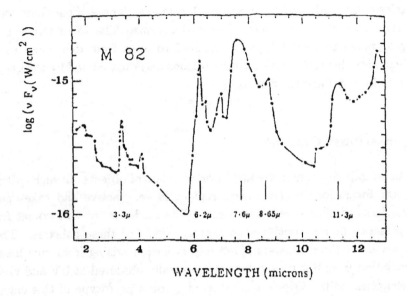

Fig. 7.9 Emission spectrum of M82 showing clearly the presence of AUIB emission features (Gillett, F.C., Kleinman, D.E., Wright, E.L. and Capps, R.W. 1975, Astrophys. J., **198**, L65; Willner, S.P., Soifer, B.T., Russel, R.W. et al. 1977, Astrophys. J., **217**, L121: reproduced by permission of the AAS).

There is evidence to show that the extinction law in M82 deviates from the extinction law seen in our Galaxy and is similar to the one seen towards the Galactic center.

7.6 Ultraluminous Infrared Galaxies

Ultraluminous Infrared Galaxies (ULIRGs) are an important constituent of our local universe. They emit more energy in the infrared region than at all other wavelengths combined. The luminosity in the region 8-1000μm could be $\geq 10^{12} L_\odot$. ULIRGs are gas rich galaxies resulting from the interaction with another galaxy and finally leading to a complete merger of the two galaxies. ULIRGs are heavily obscured by dust and the observed infrared luminosity as obtained from IRAS and ISO measurements is the reemitted radiation by the dust. Because ULIRGs are dust rich the extinction is quite high. The estimated dust extinction in ULIRGs vary between $A_v \sim$5 to 50.

The ISO-SWS data of Arp 220 indicate a value of $A_v \sim 50$. The presence of such high extinction can have important effect in shaping the spectra of these galaxies.

ULIRGs have characteristic features of both seyferts and starburst galaxies. The observations in the IR, mm and radio characteristic of ULIRGs are similar to those of starburst galaxies while the presence of nuclear optical emission line is the characteristic feature of seyfert galaxies. Some ULIRGs contain compact central radio source and highly absorbed X-ray sources indicative of active galactic nuclei (AGNs).

In the mid-infrared observation, the features at 6.2, 7.7, 8.6 and 11.3μm have been detected (AUIBs). The 7.7μm feature is quite strong. These features can be detected in much fainter and more distant sources for redshift, $z \sim 0.4$ or so. A comparison of the luminosity in AUIBs to the total far infrared luminosity (in 60+100μm) indicate a value which is roughly half the value seen in starbursts. This suggests that roughly half of the luminosity in ULIRGs came from star formation as many of the ULRIGs observed with ISO has been identified as starbursts. The silicate feature at 9.7μm has also been seen.

7.7 Merging Galaxies

Merging galaxies are a class of objects where galaxies interact and merge with each other. They result in a wide range of morphologies including bridges between the interacting galaxies and tidal tails. The interacting galaxy systems Arp 299C is a galaxy of this type. This appears to arise out of interaction of two merging galaxies IC 694 and NGC 3690. It has a large far- infrared luminosity, L (FIR) $\sim 5 \times 10^{11} L_\odot$. The physical conditions in Arp 299C is quite extreme. Molecular observations have shown the large quantities of molecular gas with densities of the order of $10^4/cm^3$.

The spectrum of Arp 299C in the spectral region 8-13μm taken from United Kingdom Infrared Telescope (UKIRT) is shown in Fig. 7.10. It reveals a number of emission features superposed on a strong continuum. The feature at 11.3μm is quite strong and a second weaker feature at 8.6μm is also present. These are the features of AUIB-type. There is no clear indication of the 10μm silicate feature. Therefore, model fit to the observed spectra was carried out to examine the presence or absence of silicate feature. It was found that the observed spectrum can be fitted well by the superposition of a continuum produced by very small grains and AUIB-like grains. Hence silicate absorption appears to be absent. The observed spectrum is found to be consistent with a luminosity source that does not produce extremely high energy photons implying the absence of a luminous AGN in Arp 299C.

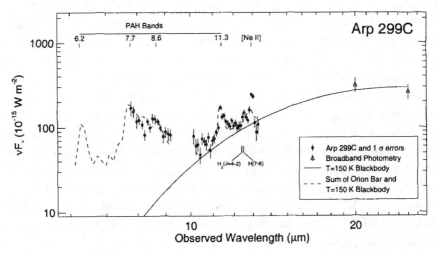

Fig. 7.10 The mid-infrared spectrum of merging galaxy Arp 299C. A hot continuum underlines the well known AUIBs emission features (Dudley, C.C. and Wynn-Williams, C.G. 1993, Astrophys. J. Lett., **407**, L65: reproduced by permission of the AAS).

7.8 Virgo and Coma Clusters

Virgo and Coma clusters are distant rich clusters of galaxies. Their distances ~20 and 100Mpc respectively. They contain galaxies of various kinds. Therefore infrared observations of these galaxies should provide valuable information on the nature of dust at farther distances.

The infrared observations of galaxies in Virgo and Coma clusters carried out with IRAS and ISO has shown that the far-infrared radiation is dominated by thermal dust emission. The results indicate that the far-infrared emission is dominated by cool dust component heated by the general interstellar radiation field compared to that of warm dust component associated with star forming regions.

The studies of interstellar medium in the Galaxy has shown that the emission in the mid-infrared region is dominated by AUIBs while emission at 15μm is contributed mostly by very small grains in regions of low radiation density. Therefore the ratio of fluxes at 6.75μm (AUIB feature) and 15μm should give an indication of the strength of the AUIB and hence could provide some information on its carrier. With this in view a large number of galaxies in both Virgo and Coma clusters have been observed with camera ISOCAM on board ISO in the two broadband passes centred at 6.75 and 15μm. The observed flux ratio of F(6.75μm) to F(15μm) for galaxies in Virgo and Coma clusters lie in the range $\sim$0.7 to 1.3. This is similar to the flux ratio in the Galaxy which varies between 1.0 to 1.8. This seems to indicate the common nature of the carrier for AUIBs in the two cases.

7.9 Quasars

Quasars are ultraluminous objects and are another enigmatic class of objects. Because they appear point-like in optical photographs these objects were called Quasi-Stellar objects abbreviated later on to Quasars. The most striking characteristic feature of quasars is their large redshifts. The most distant quasar known at the present time has a redshift of z=6.42. In terms of Big Bang cosmolgy this redshift corresponds to times when the universe was just $\sim$1 Gyr old. Therefore these objects are very useful for probing gas and dust in the early universe. Hence the search for quasars at higher and higher redshifts is an important area of study.

Quasars emit enormous amount of radiation in the far-infrared region. The study of this radiation should provide valuable information on dust.

The spectra of quasars are dominated by both emission and absorption lines. The study of emission lines of quasars provide information on the physical state and chemical composition of the gas and dust close to quasars. On the other hand the study of absorption lines with increasing redshift samples the intervening gas and dust material at larger and larger distances.

7.9.1 *Far-Infrared Radiation*

Quasars emit large amount of infrared radiation from dust was known from earlier ground-based broadband observations. This has been supported by later observations carried out with satellites. A large number of quasars have been observed with ISO in the wavelength region 5-200μm and upto redshifts z$\sim$5. The observed spectral energy distribution of some quasars at several redshifts is shown in Fig. 7.11. As can be seen from Fig. 7.11, the energy distribution has a peak around 60-100μm and decreases longward of 100μm. This clearly show that in these objects the far-infrared radiation

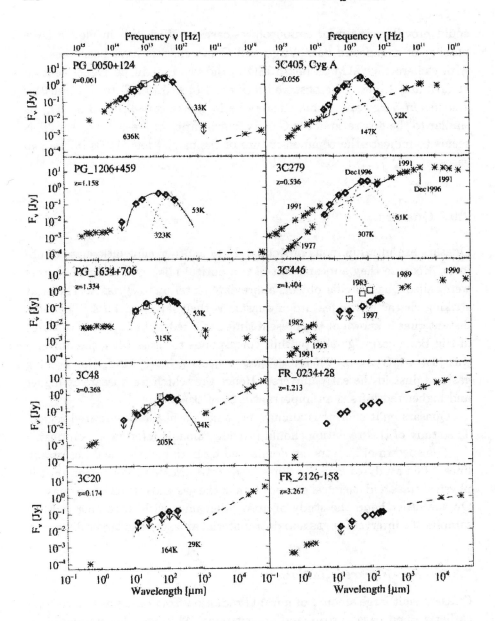

Fig. 7.11 The far-infrared emission from Quasars and Radio galaxies observed with ISO (diamonds) and IRAS (squares). The dashed lines indicate the synchrotron spectra. Several black body fit to the data is shown with temperature marked (Haas, M., Chini, R., Meisenheimer, K. et al. 1998, Astrophys. J., **503**, L109: reproduced by permission of the AAS).

is dominated by thermal emission from grains. The observed far-infrared emission from Quasars FR0234+28 and FR2126-158 is a continuation of the radio synchrotron spectra. The observations can be modelled by a superposition of several modified black bodies (i.e $F_{em} \propto B(\lambda, T_g)\lambda^{-2}$; dust emissivity $\propto \lambda^{-2}$) indicating the presence of wide variety of temperatures from hot (1000K) to cool (30K) dust in these objects. The temperature of the coolest dust component lies in the range 30 to 120K. The temperature of the cold dust component is found to vary between 30-40K at z=0 and 90 to 120K at z=2. The large variation in the grain temperature between 1000 to 30K required to fit the infrared observations indicate that the heating of the grain is by a central source with temperature decreasing with distance from the centre. The infrared luminosity at high redshift quasars upto z~5 range from 10^{46}-10^{49} ergs/sec ($\sim 10^{13}$-$10^{16} L_\odot$) and at low redshift z~0.08 the infrared luminosity $\sim 10^{45}$ ergs/sec ($\sim 10^{12} L_\odot$). The observed infrared luminosity at high redshifts is about two orders of magnitude larger than those of ultraluminous infrared galaxies.

The most distant quasar J1148+5251 at redshift z=6.42 has been observed at 250GHz (1.2mm) with IRAM 30-metre telescope and with VLA at 43GHz. The observations indicate a steeply rising spectra indicative of the thermal emission from the dust. The estimated far-infrared luminosity $\sim 10^{13} L_\odot$ and the mass $\sim 10^8 M_\odot$. Therefore the presence of large amount of dust in the highest redshift quasar at z=6.42 indicate that dust formation must have been very efficient during earlier times.

7.9.2 Spectral Studies

Quasars are very rich in their spectra with strong absorption and emission lines of elements. Therefore the study of elemental abundances from quasars should help in tracing the environment of heavy elements in the universe from early times to the present time. This in turn should provide information on the dust.

The spectra of quasars at redshifts z~6 seem to indicate that the cosmic re-ionization (i.e. from completely neutral to completely ionized state) must have occurred not too far from this type of redshift. This is based on Gunn-Peterson effect in which the trough arising due to optically thick absorption of neutral hydrogen is looked for at $\lambda_{rest} < 1216$Å. This arises as a consequence of the fact that even a tiny neutral hydrogen fraction in the intergalactic medium ($x \leq 10^{-6}$) would produce large optical depth ($\tau < 0.5$) at these wavelengths. The observation of the redshift object

J1148+5251 (z=6.42) show a sharp drop in flux blueward of the Lyman α emission and the average transmitted flux of the absorption trough is consistent with zero flux. However studies carried out with Wilkinsons Microwave Anisotropy Probe (WMAP) indicated large optical depth to Thomson scattering (τ_e=0.17±0.04) suggesting that the universe was reionized at higher redshifts, z_{ion}=17±5. These redshifts correspond roughly to time interval ~1Gyr. The re-ionization era is when the metals (and dust) are formed from the first generation stars which enrich the surrounding material. Therefore the study of metals at higher and higher redshift should also give information on the nature of dust at these redshifts.

7.9.2.1 Quasar Environment

The emission lines are very prominent in the spectra of quasars. As an illustration the spectra of a quasar is shown in Fig. 7.12. As can be seen from this figure, emission lines arise from elements such as H (Lyα), He, C, O, N, Si etc and from several ionized states. Several intercombination (semi-forbidden) lines are also present in the spectra.

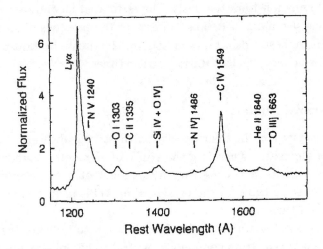

Fig. 7.12 Normalized mean spectrum of 13 quasars at z>4 showing the presence of various emission lines (Shields, J.C., Hamann, F., Foltz, C.B. and Chaffee, F.H. 1997, In *Emission Lines in Active Galaxies*, Eds. B.M. Peterson, F-Z. Cheng and S. Wilson, By the kind permission of the Astronomical Society of the Pacific Conference Series, Vol.113, p.118).

The broad emission lines are generally used for the abundance determination as it is easier to study in all quasars at all redshifts. The broad emission lines are believed to arise from clouds photoionized by the central source of a quasar. This is known as Broad Emission Line Region (BELR). The evidence for this comes from the fact that the shape of the spectra change with changes in the continuum flux. Thus the shape of the ionizing spectra is a fundamental parameter. The emission lines from photoionized clouds are determined by photoionization and emission process. Theoretical studies have been carried out which treat the ionization, temperature, radiation transfer and line emission in realistic BELR clouds in a self consistent manner. From these calculations the expected line fluxes of elements have been calculated. Comparison of these results with the observed line fluxes yield realistic abundance of elements. These studies have shown that the metallicities are typically near or a few times above the solar value. i.e. typically $Z \sim 1$ to $3Z_\odot$ upto redshift ~ 5.

The elemental abundances can also be determined from the Narrow Absorption Lines (NALs) seen from quasars. The precise location of the absorbing gas is not known but it could not be too far from the quasar. The abundances of elements determined from these lines are in good agreement with the results derived from emission line studies. Therfore the gas-phase metallicities near quasars are typically ~ 1 to $3Z_\odot$. Therefore the metal rich stars (with dust) which pollute the environment must have formed prior to the observed quasar epochs. These results are consistent with both observational and theoretical simulations in showing that the abundances at any time in the universe depends on the local density. Since quasars reside inside a higher dusty region of the centre of the galaxy at every epoch, it should have higher metallicities. Hence the dependence of metallicities on density is far more important than their ages (redshifts).

These studies clearly show that dust must be present in abundance in quasars at every redshift. Therefore it is no surprise that quasars are extremelly bright in the far-infrared region at all redshifts due to thermal emission from grains.

7.9.2.2 Lyman α Forest

The Lyman α Forest referring to discrete Lyman α lines were detected in 1970's. They consist of many hundreds of individual Lyman α absorption lines. They pervade the region of the intergalactic medium and is seen as absorption bluewards of the Lyman α emission of quasars. Because of its ubiquity and spatial correlation populations it is believed to be associated with intergalactic population rather than being associated with individual galaxies. In view of this it is interesting to determine the metal content and hence the dust content existing in these regions. The study has been made possible with the detection of absorption lines of metals associated with many of the Lyman α forest clouds, with N (HI) $\geq 10^{14}$cm^{-2}. These lines are CIV 1548, 1550Å doublet and SiIV 1394, 1403Å. There is evidence for the presence of OVI 1032, 1038Å absorption lines. The detection of such highly ionized metal lines indicate that the intergalactic medium is in a highly ionized state. The study of several quasars including the best quality spectra of quasar Q1422+231 with z_{em}=3.625 indicate carbon abundance to be 1/300 of the solar value. The study of the column density distribution of CIV abundances over the redshift interval 1.5 to 5.0 is nearly flat. Therefore the intergalactic medium was enriched in metals and dust at z>5 and since then not much change has taken place. This corresponds to a time period ~1.2 to 4.5Gyr after the Big Bang.

7.9.2.3 Damped Lyman α Systems

Damped Lyman α Systems (DLAS) are gaseous systems that are detected through their absorption lines in the optical spectra of quasars with hydrogen column density, N (HI)$\geq 2 \times 10^{20}$ cm^{-2}. Due to high column density of neutral hydrogen Damped Lyman α systems have strong Lyman α damping wings.

The dust studies in Damped Lyman α systems is generally carried out with resonance absorption lines of ZnII $\lambda\lambda$ 2026, 2062Å and CrII $\lambda\lambda$ 2056, 2062, 2066Å. These lines are selected as they are unsaturated and also Zn and Cr appears to follow closely the abundance of Fe. In addition in the local interstellar medium Zn is not incorporated into dust grains. Hence Zn should be present in the gas phase in near-solar proportions relative to H. However Cr is among the most heavily depleted elements with around 1% remaining in the gas. Also H is mostly in the atomic form. Therefore the ratio $N(Zn^+)/N(H)$ should measure directly the abundance of iron-peak elements (i.e metal enrichment) in the Damped Lyman α systems, while $N(Cr^+)/N(Zn^+)$ should give a measure of depletion of refractory elements on to dust. Taken together they should give an indication of the dust-to-gas ratio. The results of these studies show that these systems ($z\geq3$) are generally metal poor with metallicities approximately 1/13 solar. The Cr/Zn ratio indicate that around 65% of Cr is depleted onto grains. These results provide evidence for the presence of dust in high redshifed galaxies.

Even upto z∼5 metals are present and hence the dust.

7.10 Intergalactic Dust

On the basis of presently favoured star formation scenario, not enough metals are observed in gas and stars of galaxies. But the hot gas in clusters of galaxies is rich in metals. This indicate that possibly the metals tied up in dust is stripped out of the galaxies by some physical processes such as radiation pressure, galactic winds and so on. Many of the dust particles may also be evaporated in the process. The rough estimate of the total mass-injection rate from a galaxy like Arp 220 could be around 2×10^7 $M_\odot \text{Mpc}^{-3}$, out of which around 25% may be metals. Therefore even though the dust particles are expected in the intergalactic medium the amount of material and its characteristics are far from clear. This should be determined in general from observations. However the methods used for study of dust in galaxies is less effective when extended to intergalactic space. An estimate of extinction by dust was made based on the observation of lower counts of faint galaxies or distant clusters in the fields of the rich proximate clusters of galaxies. These counts required a visual extinction of the order of $A_v=0.3$ to 0.6 mag. However the detection of thermal emission from dust from Coma cluster with photometer ISOPHOT, on board ISO at 120μm gives a value for the corresponding visual extinction of $A_v=0.05$ mag, much smaller than the value quoted earlier. The limit on the intergalactic extinction has also come from the studies of the redshift evolution of the mean quasar spectral index. They indicate A_v (z=1) $\leq$ 0.05 mag. In such studies generally the standard relation between the reddening and extinction derived from galactic studies is used which use graphite and silicate dust in the form of small spheres with a distribution in radii. However it is not clear whether the same kind of dust and size distribution is applicable to intergalactic dust.

The observations of Supernova SNe Ia have long been used as a good indicator of distance in cosmology. This comes about due to the fact that the observed peak luminosity of distant SNe Ia can be normalized by their light-curve shape. The observational proof for the standard character of these objects comes from the Hubble diagram which show a nearly linear cosmic expansion of nearby SNe Ia'S. However the recent observations have shown that there is progressive dimming of the light output of Type Ia supernovae with redshift. The above observation can in principle be explained due to the presence of a grey extinction of dust particles in the intergalactic medium. In the extragalactic dust studies the interstellar two component dust model, i.e. graphite and silicate with sizes of the form N (a)$\propto a^{-3.5}$ is generally used due to lack of precise information. The size range covered is $0.005 \leq a \leq 0.25 \mu m$. The results of extinction calculations for graphite particles for sizes upto $a_{min} = 0.1 \mu m$ is found to be quite flat upto $\lambda \sim 1 \mu m$. i.e. grey extinction without reddening. When graphite and silicate dust are considered together the reddening falls by 50% for $a_{min} \leq 0.09 \mu m$. Therefore if the size of the particles is such that $a_{min} \leq 0.1 \mu m$, the grey extinction present can cause the source to become fainter with increase in distance without causing reddening. It is rather difficult to explain on the basis of dust hypothesis the observation that the distant SNe Ia'S are bluer than the local sample. However it is not clear at the present time whether this observation could be due to some other cause.

The studies of CIV absorption lines associated with Lyman α forest which is believed to arise from the low density intergalactic medium has shown the enrichment of metals (by mass) to the extent of around 1/300 solar metallicity for redshifts $z \leq 5$. This also indicate the likely presence of dust in intergalactic medium.

7.11 Intracluster Medium

The detection of iron line emission in the X-ray spectra of clusters of galaxies in late 1970s showed clearly that the intracluster gas contains a significant fraction of the gas processed in stars. The results of study on a large sample of clusters showed that the Fe abundance has a peak distribution around a mean value ~ 0.3 solar value (Fig. 7.13). Here the solar abundance of Fe is defined as log(Fe/H)=-4.33 by number. This is based on the old photospheric value for the solar Fe abundance, whereas the commonly accepted 'meteoritic' value is significantly lower (Fe/H $\sim 3.24 \times 10^{-5}$). Thus essentially all the Fe measurements in the X-ray literature should be increased by a factor of ~ 1.44 to renormalise to the meteoritic value. The spatially resolved spectra of clusters show that in general the Fe abundance is relatively uniform out to radii of $\sim$ 1Mpc. The mass of the gas in the intracluster medium (ICM) exceeds the stellar mass in the cluster galaxies by factors of 2 to 10. The ratio between them also increases with cluster richness. This indicates that the total mass of Fe is quite large. Based on the data on a large number of clusters a correlation study was

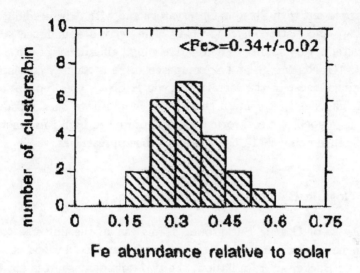

Fig. 7.13 The distribution of Fe for rich clusters of galaxies derived from Ginga data (Yamashita, K. 1992, In *Frontiers of X-ray Astronomy*, Eds. Y. Tanaka and K. Koyama, Tokyo: Universal Academy, p.475; Mushotzky, R., Loewenstein. M., Arnaud, K.A. et al. 1996, Astrophys.J., **466**, 686: reproduced by permission of the AAS).

carried out between gas mass, iron mass and optical luminosity of clusters of galaxies with the morphological type of the galaxy. For this purpose the gas mass has been deduced from X-ray imaging data on clusters with Einstein satellite and the stellar mass from the visible luminosity of the cluster of galaxies for a given M/L ratio. The total iron mass in the gas can be derived from the gas mass and for this purpose observations carried out with X-ray satellites EXOSAT and GINGA have been used to derive iron abundance. The results show that the gas mass is correlated with the stellar mass in ellipticals. This leads to the conclusion that the metals in the intracluster medium originated from elliptical galaxies and that the metals (dust) were driven out into the ICM by supernovae-driven winds.

The ratio, Mg/Fe is of interest from the point of view of onset of star formation. This comes about due to the fact that Mg comes from supernova Type IIa (α elements), while Fe comes mainly from supernova of Type Ia. The latters contribution has a built in time delay due to the large evolutionary time scale involved for the formation of white dwarf. Therefore the ratio of Mg/Fe can help in constraining the ages of star forming environment. Therefore the elemental abundances of O, Ne, Mg, Si, S, Ca, Ar and Fe for clusters of galaxies have been determined from the X-ray observations carried out with Einstein Observatory and with Advanced Satellite for Cosmology and Astrophysics (ASCA). The derived abundance pattern in each cluster is very similar. The relatively high abundance of the α-burning products (O, Ne, Mg, Si and S) compared to Fe is roughly consistent with the origin of most of the metals in Type II SNs. The implication being that the galaxies that contain these stars should have been ultraluminous during the epoch of metal formation and produced high total metal mass. The inference being that dust must have been present as well.

7.12 Cosmic Background Radiation

The spectra of Cosmic Background Radiation in the universe covers the wavelength region from γ-rays to radio. Among the various components of Cosmic Background Radiation, the two dominant sources are the Cosmic Microwave Background Radiation (CMBR) and the Cosmic Infrared Background Radiation (CIBR). CMBR is the fossil black body radiation from the hot dense plasma (Big Bang). The peak of the energy distribution occurs in the millimeter region. On the other hand, CIBR is the integrated measure of cosmic activity summed over time and space of all absorption and emission of photons, since the time of recombination ($z \leq 1000$), the time when radiation and matter got decoupled. In particular cosmic expansion will shift significant part of re-radiation of dust grains in the infrared into the wavelength regime of infrared background radiation of around 1 to 1000μm. The infrared background reflects the history of the universe from the time the first stars were formed. The CMBR is the dominant component compared to CIBR.

7.12.1 *Cosmic Infrared Background Radiation*

The Cosmic Infrared Background Radiation (CIBR) is the integrated emission of all energy sources in the universe following the decoupling of matter and radiation i.e recombination epoch. The sources contributing to CIBR include condensation of luminous objects from primardial neutral matter. In a dust free universe the CIBR can essentially be derived from a knowledge of the spectrum of the emitting sources and the cosmic history of their energy release. In a dusty universe, the total intensity of the CIBR is not changed but rather the energy is redistributed over the entire X-ray to far-infrared region of the spectrum. The prediction of the re-distributed energy distribution is a difficult task as it involves a large number of factors. They include among others, spatial distribution of sources and their cosmic history, history of dust formation and destruction, their properties etc. In spite of complex nature of the problem, several theoretical studies with varying degrees of sophistication and complexity have been carried out which takes into account in a reasonable and consistent way the energy emitted by various stellar generations at differnt metallicities, the opacity of the enriched interstellar gas and the re-radiated flux in the far-infrared and radio wavelengths. These studies had indicated that the re-radiated energy at longer wavelengths by the dust particles should be appreciable.

The direct detection of CIBR is difficult because of its faintness. The detection is also made difficult as there is no distinctive spectral signature. The only signature of the presence of this radiation is their isotropic nature on large scales. The observational effort is also made difficult because of the presence of various other contributions to the infrared sky brightness at all wavelengths which has to be corrected. Major sources of uncertainty refers to the stellar background (1.2 to 3.5μm), the interplanetary dust (1.25 to 100μm) and the interstellar dust (100 to 240μm). Among these the interplanetary dust contribution is the most difficult problem to handle since it dominates the measured sky brightness from 1 to 140μm even at high galactic and ecliptic latitudes.

The instruments on COBE mission was designed to search for the CIBR. It had two instruments for making absolute sky brightness measurements. They are the Diffuse Infrared Background Experiment (DIRBE) and the Far Infrared Absolute Spectrometer (FIRAS). The DIRBE instrument was an absolute photometer with 10 broad photometric bands at 1.25, 2.2, 3.5, 4.9, 12, 25, 60, 100, 140 and 240μm. This instrument was designed primarily to search for the CIBR from 1.25 to 240μm. The FIRAS instrument was a Fourier transformer spectrometer for measuring sky brightness at wavelengths from 100μm to 1cm. This instrument designed primarily to make a definitive measurement of the CMBR spectrum and to extend the search for the CIBR to millimeter wavelengths. Therefore DIRBE and FIRAS instruments provided high quality photometric data to search for CIBR. After extensive modeling and removing the contributions from various sources there still remained a substantial background energy in the far-infrared and sub-millimeter with a comparable level in the infrared. This clearly showed the presence of CIBR. Assuming that star formation is the major source of the observed background the large amount of energy in the long wavelength background suggests that substantial amount of star formation activity at early times was embedded in dust. The present observations indicate that the energy density in the 0.3 - 2000μm background is about 10% of that in the cosmic microwave background.

7.12.2 Cosmic Microwave Background Radiation

Cosmic Microwave Background Radiation (CMBR) refers to the uniform background radiation in the universe with a blackbody temperature of about 2.7K. This radiation was detected in 1965 by radio means. The presence of this radiation and its precise spectrum is of primary importance to Cosmology. The measurements carried out with the COBE satellite has established that the observed microwave radiation is an extremelly smooth curve and fits extremelly well with a blackbody spectrum corresponding to a temperature of 2.726K. The CMBR is generally interpreted as relic radiation from a hot dense phase of the universe. In this interpretation, as the universe expands the intensity of the radiation would diminish but the remnant of this primardial radiation should still be detectable today as the faint background of microwave radiation. Therefore the detection of CMBR is taken as a support for the Big Bang Cosmology. An alternative explanation for the CMBR has been proposed based on thermalization by grains in a Quasi Steady State Cosmology (QSSC) model of the universe. In QSSC, matter is created in each cycle at a modest rate throughout the universe. However, the radiation background is maintained but the energy density is expected to fall from one cycle to the next. The decrease in the background radiation energy density is compensated by the thermalization of starlight produced during each cycle. The estimated energy density of CMBR at the present time from such a process appears to be consistent with the observed value $\sim 4 \times 10^{-13}$ergs/cm^3, corresponding to T$\sim$2.7K. In this scenario, thermalization of starlight is crucial. It is found that iron wiskers of sizes $\sim$0.5-1mm with their wavelength dependent extinction property can thermalize the starlight. Such particles are produced in the neighbourhood of supernovae and ejected out of the galaxy through shock waves etc. The estimated density $\sim 10^{-35}$gm/cm^3 of such wisker particles present at the time of oscillatory phase is quite sufficient for thermalization of starlight.

In brief, the infrared observations of galaxies seems to indicate the grain material to be composed of silicate and carbonaceous material. The aromatic unidentified infrared bands are weaker in active galaxies, where the environment is severe, implying the destruction of the carrier. Large Magellanic Cloud has a weaker 2175Å feature in the extinction curve and is absent in Small Magellanic Cloud. Therefore the carrier responsible for 2175Å feature is less in LMC compared to Galaxy and more so in SMC. Both amorphous and crystalline silicates are present in LMC. The presence of AUIB-carrier is indicated in Virgo and Coma clusters. The intracluster medium is rich in iron came from the study of iron line emission in the X-ray region. This could have come from ellipticals by supernovae driven winds implying that dust must be present in these objects. Dust is quite abundant in quasars even at redshift $z\sim 6.4$. The study of metallic lines in the absorption spectra of quasars at high redshifts which samples the intervening gas, indicate that refractory elements are depleted from the gas phase. Hence they are tied up in grains. These high redshifts correspond roughly to the evolution time scale of just ≤ 1 Gega year or so depending on cosmological models. The detection of Cosmic Infrared Background Radiation suggests that substantial amount of star formation activity at early times was embedded in dust. Therefore dust appears to be present in the entire universe.

References

Good account of IR observations is given in the following references
Genzel, R. and Cesarsky, C.J. 2000, Ann. Rev. Astron. Astrophys., **38**, 761.
Sanders, D.B. and Mirabel, I.F. 1996, Ann. Rev. Astron. Astrophys., **34**, 749.
Soifer, B.T., Houck. J.R. and Neugebauer, G. 1987, Ann. Rev. Astron. Astrophys., **25**, 187.

Normal Galaxies
Alton, P.B., Trewhella, M., Davis, J.I., Evans, R., Bianchi, S., Gear, W., Thronson, H., Valentijn, E. and Witt, A. 1998, Astron. Astrophys., **335**, 807.
Dale, D.A., Silbermann, A., Helou, G., Valjavec, E., Malhotra, S. et al. 2000, Astron. J., **120**, 583.
Helou, G., Lu, N.Y., Werner, M.W., Malhotra, S. and Silbermann, N. 2000, Astrophys. J., **532**, L21.
Hunter, D.A., Kaufman, M., Hollenbach., D.J., Rubin, R.H., Malhotra, S., Dale, D.A., Brauher, J.R., Silbermann, N.A., Helou, G., Contursi, A. and Lord. D. 2001, Astrophys. J., **553**, 121.
Lu, N., Helou, G., Werner, M.W., Dinerstein, H.L., Dale, D.A., Silbermann, N.A., Malhotra, S., Beichman, C. and Jarrett, T.H. 2003, Astrophys. J., **588**, 199.
Sauvage, M. 1997, In *Interstellar Medium in Galaxies*, Ed. J.M. van der Hulst, Kulwer Academic Publishers, p.1.

SMC

Prevot, M.L., Lequeux, J., Maurice, E., Prevot, L. and Rocca-Volmerange, B. 1984, Astron. Astrophys., **132**, 389.
Reach, W.T., Boulangor, F., Contursi, A. and Lequeux, J. 2000, Astron. Astrophys., **361**, 895.
Rodrigues, C.V., Magalhaes, A.M, Coyne, G.V. and Pirola, V. 1997, Astrophys. J., **485**, 618.

LMC

Clayton, G.C., Martin, P.G. and Thomson, I. 1983, Astrophys. J., **265**, 194.

Silicate feature in LMC

Glass, I.S. 1984, Mon. Not. Roy. Astron. Soc., **209**, 759.
Voors, R.H.M., Waters, L.B.F.M., Morris, P.W., Trams, N.R., Koter, A.de. and Bouwman, J. 1999, Astron. Astrophys., **341**, L67.

Seyfert galaxies

Clevel, J., Schulz, B., Altieri, B., Barr, P., Claes, P., Heras, A., Leech, K., Metcalfe, A. and Salama, A. 2000, Astron. Astrophys., **357**, 839.
Perez Garcia, A.M., Rodriguez Espinosa, J.M. and Santolaya Roy, A.E. 1998, Astrophys. J., **500**, 685.

Starburst galaxies

Forster Schreiber, N.M., Genzel, R., Lutz., D., Kunze, D. and Sternberg, A. 2001, Astrophys. J., **552**, 544.
Gillett, F.C., Kleinmann, D.E., Wright, E.L. and Capps, R.W. 1975, Astrophys. J., **198**, L65.
Sturm, E., Lutz, D., Tran, D., Feuchtgruber, H., Genzel, R., Kunze, D., Moorwood, A.F.M. and Thornley, M.D. 2000, Astron. Astrophys., **358**, 481.
Weedman, D.W., Feldman, F.R., Balzano, V.A., Ramsey, L.W., Sramek, R.A. and Wu, Chi-Cho. 1981, Astrophys. J., **248**, 105.
Willner, S.P., Soifer, B.T., Russel, R.W., Joyce, R.R. and Gillett, F.C. 1977, Astrophys. J., **217**, L121.

ULIRG

Genzel, R., Lutz, D., Sturm, E., Egami, E., Kunze, D., Moorwood, A.F.M., Rigopoulou, D., Spoon, H.W.W., Sternberg, A. et al. 1998, Astrophys. J., **498**, 579.
Tran, Q.D., Lutz, D., Genzel, R., Rigopoulou, D., Spoon, H.W.W., Sturm, E., Gerin, M., Hines, D.C., Moorwood, A.F.M., Sanders, D.B., Scoville, N., Taniguchi, Y. and Ward, M. 2001, Astrophys. J., **552**, 527.

Merging Galaxies

Dudley, C.C. and Wynn-Williams, C.G. 1993, Astrophys. J., **407**, L65.

Clusters

Arnaud, M., Rothenflug, R., Boulade, O., Vigroux, L. and Vangioni-Flam, E. 1992, Astron. Astrophys., **254**, 49.

Bicay, M.D. and Giovanelli, R. 1987, Astrophys. J., **321**, 645.

Boselli, A., Lequeux, J., Contursi, A., Gavazzi, G., Boulade. O., Boulanger, F., Cesarsky, D., Dupraz, C., Madden, S., Sauvage, M., Viallefond, F. and Vigroux, F. 1997, Astron. Astrophys., **324**, L13.

Mitchell, R.J., Culhane, J.L., Davison, P.J. and Ives, J.C. 1976, Mon. Not. Roy. Astron. Soc., **176**, 29.

Mushotzky, R., Loewenstein, M., Arnaud, C.A., Tamura, T., Fukazawa, Y., Matsushita, K., Kikuchi, K. and Hatsukade, I. 1996, Astrophys. J., **466**, 686.

Reionization

Becker, R.H., Fan, X., White, R.L., Strauss, M.A., Narayanan, V.K., Lupton, R.H., Gunn, J.E. et al. 2001, Astron. J., **122**, 2850.

Fan, X., Strauss, M.A., Schneider, D.P., Becker, R.H., White, R.L. et al. 2003, Astron. J., **125**, 1649.

Gnedin, N.Y. and Prada, F. 2004, Astrophys. J. Lett., **608**, L77.

Loeb, A and Barkana, R. 2001, Ann. Rev. Astron. Astrophys., **39**, 19.

Quasars
Infrared Emission

Bertoldi, F., Carilli, C.L., Cox, P., Fan, X., Strauss, M.A., Beelen, A., Omont, A. and Zylka, R. 2003, Astron. Astrophys., **406**, L55.

Haas, M., Muller, S.A.H., Chini, R., Meisenheimer, K., Klaas, U., Lemke, D., Kreysa, E. and Camenzind, M. 2000, Astron. Astrophys., **354**, 453.

Wilkes, B.J., Hooper, E.J., McLeod, K.K., Elvis, M.S., Impey, C.D., Lonsdale, C.J.,Malkan, M.A. and McDowell, J.C. 1999, *The Universe as seen by ISO*, Eds. P. Cox and M.F. Kessler, ESA SP-427, p.845.

Emission lines

Hamann. F. and Ferland, 2000, Ann. Rev. Astron. Astrophys., **38**, 487.

Hamann, F., Korista, K.T., Ferland, G.J., Warner, C. and Baldwin, J. 2002, Astrophys. J., **564**, 592.

Warner, C., Hamann, F., Shields, J.C., Constantin, A., Foltz, C.B. and Chaffee, H. 2002, Astrophys. J., **567**, 68.

Lyman α Forest

Cowie, L.L., Songaila, A., Kim, T.S. and Hu, E.M. 1995, Astron. J., **109**, 1522.

Pettini, M., Madau, P., Bolte, M., Prochaska, J.X., Ellison, S.L. and Fan, X. 2003, Astrophys. J., **594**, 695.

Songaila, A. 2001, Astrophys. J., **561**, L153.

Damped Lyman α Systems

Hou, J.L., Boissier, S.and Prantzos, N. 2001, Astron. Astrophys., **370**, 23.

Nestor, D.B., Rao, S.M., Turnshek, D.A. and Berk, D.V. 2003, Astrophys. J., **595**, L5.
Pettini, M., Smith, L.J., Hunstead, R.W. and King, D.L. 1994, Astrophys. J., **426**, 79.
Pettini, M., Smith, L.J., King, D.L. and Hunstead, R.W. 1997, Astrophys. J., **486**, 665.
Prochaska, J,X., Gawiser, E., Wolfe, A.M., Castro, S. and Djorgovski, S.G. 2003, Astrophys. J., **595**, L9.

Intergalactic dust, Supernova
Aguirre, A. 1999, Astrophy. J., **25**, 583.
Leibundgut, B. 2001, Ann. Rev. Astron. Astrophys., **39**, 67.

Cosmic Infrared Background Radiation
Franceschini, A., Mazzel, P., De Zotti, G. and Danese, L. 1994, Astrophys. J., **427**, 140.
Hauser, M.G. and Dwek, E. 2001, Ann. Rev. Astron. Astrophys., **39**, 249.
Puget, J.-L., Abergel, A., Bernard, J.-P., Boulanger, F., Burton, W.B., Desert, F.-X. and Hartmann, D. 1996, Astron. Astrophys., **308**, L5.

Cosmic Microwave Background Radiation
Narlikar, J.V and Padmanabhan, T. 2001, Ann. Rev. Astron. Astrophys., **39**, 211.
Penzias, A.A. and Wilson, R.W. 1965. Astrophys. J., **142**, 419.

Dust Formation in the Early Universe
Nozawa, T., Kozasa, T., Umeda, H., Maeda, K. and Nomoto, K. 2003, Astrophys. J., **598**, 785.
Schneider, R., Ferrara, A. and Salvaterra, R. 2004, Mon. Not. Roy. Astron. Soc., **351**, 1379.

Chapter 8

Epilogue

Nature and composition of dust in the Universe is derived from astronomical observations and theoretical studies suplemented by laboratory studies of dust analogues. These aspects have been discussed in previous chapters. Nature of dust have been derived by various techniques. Some of the salient and important results resulting from these studies may be summarised as follows.

A strong and broad absorption feature centered around 2175Å has been detected in the interstellar extinction curve. This is attributed to some type of carbonaceous material and in particular graphite.

A group of strong emission lines centered at 3.3, 6.2, 7.7, 8.6 and 11.3μm have been detected from many sources in the Galaxy. They are generally attributed to PAHs in some form.

The two strong and broad absorption or emission features centred at 9.7 and 19μm in the spectra of different kinds of objects such as, O-star, HII regions, Galactic nuclei etc. are attributed to silicate dust. The silicate dust which is mostly crystalline, Magnesium-rich and of complex mineralogy, comes from the presence of a large number of structural features in the spectra.

Studies of the Galaxy have shown that photoluminescence can also give information on the dust particles. In particular they are attributed to nano-silicon and nano-graphite particles.

Extraction of dust particles of graphite, diamond, Magnesium-rich oxide, silicon carbide, FeS etc. from meteorites and carbonaceous and silicate type of grains in interplanetary dust particles and comets have given vital information on the nature of grains in the Galaxy. The exact similarity of the dust composition, forsterite, between Comet Hale-Bopp and Herbig Ae/Be star is striking.

Stardust mission will provide valuable information for the solution of several basic problems pertaining to meteorites, interplanetary dust particles, comets and origin of life on Earth etc. The composition of collected interstellar grains will provide direct proof for the growth of grain in circumstellar outflows some of which will survive in the interstellar medium before getting incorporated in meteorites and entering the solar system.

Outside our Milky Way the silicate and carbonaceous type of grains are common in normal and in active galaxies. Farther in space, carbonaceous type of dust grains with the added presence of silicate are indicated in Virgo and Coma cluster of galaxies

Molecular equilibrium calculations as well as isotopic spectral features indicate that carbonaceous type of grains (graphite, diamond, SiC, PAHs) are the product of carbon-rich stars and silicate grains that of oxygen-rich stars.

Techniques have improved enormously that we are able to infer the nature of dust at a time close to 1Gyr using the observations of quasar at redshift of z=6.42. This shows that dust formation is efficient right from the beginning. Quasars are also found to be metal-rich with metallicities of solar or a few times above the solar value indicating that substantial star formation must have taken place prior to the observed quasar epochs producing large amount of dust. The study of absorption lines of elements in quasars upto redshift $z \sim 5$ or so, arising from the intervening material has shown the presence of metals and hence dust.

Detection of Infrared Background Radiation arising mainly from thermal re-radiation from dust grains indicate the presence of dust at earlier epochs.

Therefore, the fundamental questions of interest are, when was the first dust formed, can dust form in a metal free environment and what was the nature of this dust and whether there is any variation in the composition of dust as a function of time.

Theoretical studies indicate the source of dust in the early universe can only be from Supernovae Type II arising from metal free objects of masses $\sim 140\text{-}260 M_\odot$, as their evolutionary time scale is shorter than Hubble time. The first solid particles in the universe may be Mg_2SiO_4 (forsterite), amorphous carbon, magnetite (Fe_3O_4) and corundum (Al_2O_3). Further, observations also indicate the exciting possibility of the presence of two re-ionization epochs separated by a relatively small redshift interval where universe might have recombined again. This should result in a complex behaviour of the composition of dust. The ultimate answers to many of these questions have to come from observations. Therefore it is essential to probe the universe at higher redshifts. James Webb Space Telescope (formerly known as Next Generation Space Telescope) expected to be launched in 2011 will answer many of these outstanding questions. We are, therefore, in the exciting period of the study of dust in the universe.

Appendix A

H-R Diagram

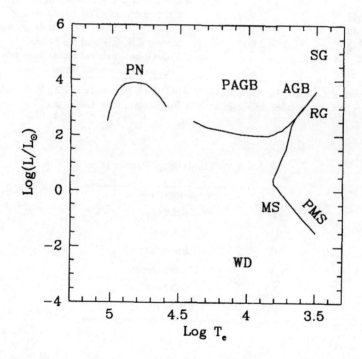

Fig. A.1 Schematic H-R diagram indicating the location of various types of stars. MS (Mainsequence), PMS (Pre-mainsequence), RG (Redgiant), SG (Supergiant), AGB (Asymptotic giant branch), PAGB (Post- asymptotic giant branch), PN (Planetary nebula) and WD (White dwarf).

Table A.1 Spectral Classification

Spectral type	Typical temperature $T_{eff}(K)$	Prominent spectral lines
O	40,000	HeII dominant, SiIV, OII also strong.
B0	28,000	HeI strong, HeII weak, CIII, SiIII, OII present, H developing in later type
A0	9,900	Strong H Balmer lines, HeI absent, MgII, SiII strong.
F0	7,400	H lines weaker, CaII lines growing strong, neutral metal lines FeI, CaI etc. growing.
G0	6,000	CaII intense, neutral metals FeI, CaI strong.
K0	4,900	CaII lines at maximum, neutral metal lines strong, molecular bands like CH and CN developing
M0	3,500	CaI prominent, TiO bands strong
M5	2,800	CaI more intense, TiO bands stronger
R,N(C)	2,800	Strong CN, CH and C_2 bands, TiO absent, neutral metal lines present
S	2,800	ZrO bands, neutral metal lines

Agrawal, P.C. 2003, In *Astrophysics: A Modern Perspective*, Ed. K.S. Krishna Swamy, New Age International Limited Publishers, New Delhi, p.1

Table A.2 Luminosity Classification

I	Supergiant
II	Bright Giant
III	Giant
IV	Sub-giant
V	Mainsequence
VI	Subdwarf
VII	White dwarf

Appendix B

Stellar Evolution

Table B.1 States of Stellar Evolution for Different masses

Mass	Evolutionary Sequence*
Low mass stars <$5M_\odot$	Protostar → pre-mainsequence → mainsequence ($1M_\odot$, $\sim 10^{10}$ yrs) → red giant ($\sim 10^9$ yrs) → Planetary nebula → white dwarf.
Intermediate mass stars $(5$-$10)M_\odot$	Protostar → mainsequence ($7M_\odot$, $\sim 3 \times 10^7$ yrs) → redgiant ($\sim 10^6$ yrs) → planetary nebula or supernova
Massive stars $(10$-$20)M_\odot$	Protostar → mainsequence ($18M_\odot$, $\sim 10^7$ yrs) → supergiant ($\sim 10^3$ - 10 yrs) → supernova

*The representative age of the star for the corresponding mass is given within brackets

Table B.2 Different stages of evolution of $7M_\odot$ star

Energy Source	Lifetime	Location in the H-R diagram
Core hydrogen burning	2.6×10^7 yrs	Mainsequence
Shell hydrogen burning	5×10^5 yrs	Subgiant redgiant branch
Core helium burning + shell hydrogen burning	1.5×10^6 yrs	Red supergiant
Double shell source	$\sim 2 \times 10^6$ yrs	Supergiant
Shell helium burning	$\sim 10^6$ yrs	Red supergiant
Explosive ignition of carbon		Star explodes as a supernova

Ramadurai, S. 2003, In *Astrophysics: A Modern Perspective*, Ed. K.S. Krishna Swamy, New Age International Limited Publishers, New Delhi, p.209

Appendix C

Nucleosynthesis

Table C.1 Summary of nucleosynthetic processes

	Process	Site	Major products
Primordial nucleosynthesis	In a universe expanding and cooling somewhat faster than the neutron half life, the right neutron/proton ratio obtains, leading to nucleosynthesis	Hot big bang	^{2}H,^{3}He, ^{4}He,^{7}Li
Hydrogen burning	p-p chain CNO cycle	Low mass stars stars with $M > 1.4 M_\odot$	^{4}He, ^{14}N
Helium burning	triple α and other helium burning reactions	Stellar interiors beyond the main sequence	^{12}C,^{16}O ^{20}Ne,^{24}Mg
Explosive hydrogen and helium burning	Hot CNO cycle with helium as fuel	External layers of novae and supernovae	^{13}C,^{15}N ^{17}O,^{18}O ^{19}F,^{21}Ne
Hydrostatic C,O and Si burning	Heavy ion thermonuclear reactions + Neutron captures on slow time scale	Stellar interiors beyond the main sequence	^{24}Mg,^{28}Si, ^{32}SNe to Ca
Explosive C,O and Si burning	C,O & Si burning temp. $\sim 2\text{-}5 \times 10^9$K and density $\sim 10^5 \text{-} 2 \times 10^7$ g.cm^{-3}	Dense regions of supernovae	^{28}Si,^{31}P, ^{32}S,^{33}S,^{34}S, ^{36}S,^{35}Cl,^{35}Ar, ^{36}Ar,^{41}Ca,^{42}Ca, ^{46}Ti,^{52}Cr,^{54}Fe, ^{56}Fe,^{58}Ni, ^{54}Mn,^{58}Fe,
Equilibrium process	Statistical equilibrium due to photo-disintegration of Fe	pre-supernovae supernovae	Fe peak nuclides
Spallation	Galactic cosmic rays interacting with interstellar CNO nuclei	Interstellar medium	Li,Be,B
r process	neutron capture on fast time scale	supernovae	unshielded isobars with $A \geq 62$
s process	neutron capture on slow time scale	orderly evolution of stellar interiors	most stable isobars with $A \geq 62$
p process	proton capture and photonuclear reactions + photonuclear reactions on slow time scale	Supernovae stellar interiors	bypassed isobars with $A \geq 62$
ν process	inelastic scattering of neutrinos off nuclei	supernovae	^{7}Li,^{11}B,^{19}F, ^{138}La,^{180}Ta

Ramadurai, S. 2003, In *Astrophysics: A Modern Perspective*, Ed. K.S. Krishna Swamy, New Age International Publishers, New Delhi, p.209.

Appendix D

Acronyms

AGB	:	Asymptotic Giant Branch
AGN	:	Active Galactic Nuclei
ALMA	:	Atacama Large Millimeter Array
ASCA	:	Advanced Satellite for Cosmology and Astrophysics (Japan)
BELR	:	Broad Emission Line Regions
Chandra	:	X-ray Observatory launched in 1999 (USA)
CHON	:	Particles Composed of H,C,N and O
CIBR	:	Cosmic Infrared Background Radiation
CMBR	:	Cosmic Microwave Background Radiation
COBE	:	Cosmic Background Explorer (Satellite)
CODAG	:	Cosmic Dust Aggregate Experiment
DIRBE	:	Diffuse Infrared Background Experiment on COBE
DLAS	:	Damped Lyman α Systems
DDA	:	Discrete Dipole Approximation
DGL	:	Diffuse Galacti Light
DIRBE	:	Diffuse Infrared Background Experiment on COBE
DLA	:	Damped Lyman α regions
		High Energy Astrophysical Observatory (HEAO-2) (Renamed Eins
ERE	:	Extended Red Emission
EXOSAT	:	European X-Ray Observatory Satellite
FIR	:	Far Infrared(30 to 300μm)
FIRAS	:	Far Infrared Absolute Spectrometer on COBE
FUV	:	Far Ultraviolet
FWHM	:	Full Width at Half Maximum Intensity
GEMS	:	Glass with Embedded Metal and Sulphides
Ginga	:	Japanese X-Ray Satellite

HAC	:	Hydrogenated Amorphous Carbon
HR	:	Hertzsprung-Russel (Diagram)
HST	:	Hubble Space Telescope
ICM	:	Intracluster Medium
IDP	:	Interplanetary Dust Particles
IGM	:	Intergalactic Medium
IRAM	:	Institute de Radio Astromica Milimetrica (Spain)
IRAS	:	Infrared Astronomical Satellite
ISM	:	Interstellar Medium
ISO	:	Infrared Space Observatory
ISOCAM	:	Camera on Board ISO
ISOPHOT	:	Photometer on Board ISO
ISO-SWS	:	The Infrared Space Observatory (ISO) Short Wavelength Spectro
ISO-LWS	:	The Infrared Space Observatory (ISO) Long Wavelength Spectror
KAO	:	Kuiper Airborne Observatory
IUE	:	International Ultraviolet Expolrer (Satellite)
LMC	:	Large Magellanic Cloud
MIR	:	Mid Infrared (5 to $30\mu m$)
MIS	:	Matrix Isolation Spectroscopy
MRN	:	Mathis, Rumpel and Nordsieck (Grain size distribution)
NAL	:	Narrow Aborption LInes
NGST	:	Next Generation Space Telescope
NIR	:	Near Infrared (1 to $5\mu m$)
PAH	:	Polycyclic Aromatic Hydrocarbons
PIA	:	Particulate Impact Analyser on Giotto Spacecraft
Planck	:	Name of the space Observatory for Mapping Microwave Backgrou Fluctuation
PN	:	Planetary Nebula
PUMA	:	Detectors on Vega Spacecraft
QCC	:	Quenched Carbonaceous Candensates
QSSC	:	Quasi Steady State Cosmology
QSO	:	Quasi Stellar Object
RGB	:	Red Giant Branch
ROSAT	:	Roentgen (X-Ray) Satellite (Germany)

SCUBA	:	Submillimeter Common User Bolometer Array
SMC	:	Small Magellanic Cloud
SOFIA	:	Stratospheric Observatory for Infrared Astronomy
SN	:	Supernova
UIR	:	Unidentified Infrared
UKIRT	:	United Kingdom Infrared Telescope
ULIRG	:	Ultraluminous Infrared Galaxy
UV	:	Ultraviolet
WMAP	:	Wilkinson Microwave Anisotropy Probe
XMM	:	X-ray Multi-Mirror Telescope (also called XMM-Newton)
YSO	:	Young Stellar Object

Index

Absorbance, 9, 12, 16, 17
Absorption crosssection, 10, 79
Active galaxies, 203, 222, 230, 236
AGB stars, 4, 29, 101, 106, 165, 169, 170, 172, 175, 178, 179, 181–185, 188, 194, 197
Aggregation, 46, 49–51, 55, 113
Albedo, 37, 74–77, 124, 125, 142, 145, 148
Aliphatic, 8, 24–26, 34, 35, 84, 98, 135, 150, 159
Amount of extinction, 58–60, 65
Annealing, 19, 21, 22, 26, 191
Apolar ices, 13, 14, 85, 88, 91, 96
Aromatic, 8, 24, 26, 27, 29, 32, 36, 52, 85, 134, 150, 209
Aromatic unidentified infrared bands, 2, 186, 230
Asteroids, 153, 154, 164
Astronomical silicate, 39
Asymmetry factor, 38
Asymptotic giant branch stars, 18
AUIBs, 32, 78, 86, 107, 134, 181, 186–188, 192–194, 196, 199, 208–213, 215–219, 231

Band strength, 9, 13, 17, 30

Capture velocity, 50
Carbon-rich stars, 2, 27, 102, 107, 176, 177, 180, 181, 195, 236
Carbonaceous chondrites, 148, 153, 158, 163, 164
Cauliflower-like structure, 51, 159
Chemical subgroups, 6, 7
CHON particles, 105, 134, 135, 137, 141, 143, 150
Chondrules, 158, 163
Circumstellar chemistry, 178
Circumstellar dust, 52, 142, 169, 182
Cirrus, 82
Coagulation, 19, 48–50, 72
Cometary dust, 105, 111, 113, 123, 135, 137–139, 141–143
Condensation, 19, 25, 44, 48, 87, 91, 160, 161, 165, 173, 177, 187, 195–200, 202, 228
Condensation temperature, 80, 173, 174, 176, 177, 186
Continuum emission, 132
Cosmic background radiation, 228
Cosmic infrared background radiation, 228, 231, 234
Cosmic microwave background radiation, 228, 230, 234
Crystalline silicate, 17, 21, 139, 183, 186, 188, 191, 193, 209
Crystallization, 21, 47

Damped Lyman α systems, 224, 233
Depletion of elements, 2, 80, 81, 199, 224
Diagnostic bands, 7
Diamond, 23, 24, 26, 39, 106, 160,

165, 189, 235, 236
Dielectric constant, 38, 39
Diffuse galactic light, 73–75, 77
Diffuse interstellar bands, 37, 54, 81, 82, 208
Dirty ices, 16, 70, 101, 142
Discrete-dipole approximation, 38
Dust at high redshifts, 221, 224
Dust formation, 102
Dust tail, 111–117, 123
Dust to gas ratio, 85
Dust trail, 116, 117, 144

Elemental depletion, 79, 108
Extended red emission, 22, 77, 108
Extinction curve, 26, 34, 40, 64, 65, 67–71, 74, 99, 206
Extragalactic dust, 203, 212, 226
Extraterrestrial origin, 153, 164

Far-infrared, 7, 82, 86, 122, 123, 181, 193, 210–213, 218–221, 223, 228, 229
Fluffy, 19, 45, 46, 49, 51, 71, 92, 105, 113, 147
Fractal aggregates, 45

Galactic centre, 84, 96–99
GEMS, 103, 106, 148, 150, 151, 153, 156, 157, 182
Giotto spacecraft, 105, 113, 130, 132, 134, 135, 137, 138, 141, 142
Graphite, 23–25, 39, 40, 70, 71, 99, 102, 106, 107, 122, 159–162, 165, 166, 175–177, 180, 181, 195, 196, 198, 225, 226, 235, 236
Gunn-Peterson effect, 221

Haro 3 galaxy, 29, 31
Herbig Ae/Be stars, 26, 130, 152, 156, 166, 187, 188, 190, 201, 235
Hetronuclear molecule, 6
HII regions, 31, 58, 67, 77, 84, 85, 108, 109
Homonuclear molecule, 6

Hydrocarbons, 8, 23, 25, 37, 52, 84, 98, 135, 150, 159
Hydrogenated amorphous carbon, 23, 25, 36, 70, 78, 84
Hydrogenation, 11, 25, 33, 78, 95

Ices, 12–16, 48, 52, 83, 88–90, 92, 94–96, 113, 142, 143
IDPs, 105–107, 128, 143, 147, 148, 153, 182, 199
Inclusions, 157, 158, 160, 161, 163, 181
Infrared emission, 113, 122
Infrared spectroscopy, 6, 13, 21
Integrated absorbance, 9, 10, 17
Integrated band strength, 10
Interplanetary dust, 19, 46, 50, 51, 103–105, 113, 147, 148, 153, 182, 199
Interstellar ices, 11, 13
Intracluster medium, 226
Isotopic studies, 141, 146, 161, 165

Kramers-Kronig relation, 39, 73

Laboratory studies, 5, 6, 19, 25, 33, 37
Large magellanic cloud, 205, 209, 231
Lyman α forest, 223

Magellanic cloud, 205
Mass absorption coefficient, 23, 41
Mass loss, 169–172
Mass spectra, 33, 162
Merging galaxies, 217
Meteorites, 51, 103, 105–107, 138, 142, 147, 148, 150, 154, 156–169, 175, 177, 181, 182, 198, 199, 235, 236
Microgravity studies, 45
Microwave scattering, 44
Mid-infrared, 7, 12, 29
Mie theory, 39, 44, 68, 69, 117, 119, 121
Milky way galaxy, 205
Mineralogical composition, 138, 139

Mixed ices, 10, 12, 53
Molecular clouds, 58, 83, 86–88, 91, 92, 100, 101, 209, 215

Nano-crystals, 22, 23, 79
Nano-particles, 78, 79, 156
Near infrared, 28
Normal galaxies, 203, 209, 212, 231
Novae, 102, 103, 165, 169, 195–197, 199, 202
Nucleation, 19, 20, 48, 49, 55, 101, 102, 159, 162, 172, 177, 198
Nucleosynthesis, 105, 160, 165, 169, 170, 199

Onion-like structure, 159, 197
Oort limit, 64, 67, 71
Optical constants, 23, 38, 39
Optical depth, 9, 10, 69, 97, 125, 215, 221, 222
Optical properties, 19, 25, 26, 37, 39
Organics, 15, 90, 99, 101, 133–135, 142, 145, 159
Oxygen-rich stars, 2, 170, 173, 177, 182, 185, 186, 236

PAHs, 2, 23–25, 27–34, 53, 78, 86, 107, 150, 162, 169, 178, 189, 235, 236
Phase function, 38, 74, 76, 77, 119, 120
Photoluminescence, 22, 23, 77–79, 156, 235
Planetary nebula, 181, 182, 192–195, 197
Plasma-dust interaction, 52
Platt particles, 70
Polar ices, 13, 14, 85, 91, 96
Polarization, 2, 40, 46, 59, 72, 73, 113, 120–122, 142, 144, 205, 208
Polyatomic molecule, 6
Polycyclic aromatic hydrocarbon, 2, 24, 27
Polycyclic aromatic hydrocarbons, 23, 70
Poynting-Robertson effect, 153

Pre-main sequence stars, 187, 188
Presolar grains, 105, 106, 141, 156, 159–162, 165, 169

Quasars, 3, 204, 219–224, 231, 233, 236

R Coronea Borealis, 27, 180
Radiation pressure, 111, 114, 115, 117
Re-ionization, 221, 222
Red giants, 171
Reddening curve, 64, 65, 67, 68, 70, 71
Reduced gravity, 45
Reduced gravity research facility, 45, 46
Reflection nebula, 58, 75–78, 85, 86
Refractive index, 39–41

S stars, 170, 177, 186
Scattering, 37, 44, 73
Seyfert galaxy, 204, 213–215, 217, 232
Silicate, 2, 17, 19–21, 23, 36, 39, 40, 42–44, 53, 71, 72, 75, 83, 85, 88, 90, 91, 93, 97, 99–101, 103, 106, 107, 116, 119, 122, 123, 126–131, 135, 137, 139, 142, 148–152, 154, 156–158, 169, 181–188, 192, 195–197, 199, 208, 210, 213, 215, 217, 218, 225, 226, 230, 232, 235, 236
Silicon carbide, 46, 103, 160, 175, 180, 181
Small magellanic cloud, 205, 207, 231
Solar flare track, 153, 154
Sources of dust, 101, 110
Space shuttle, 46, 47
Spacecraft studies, 103
Spectral features, 113, 122, 126–128, 135
Spectral studies, 221
Starburst galaxy, 210, 215–217, 232
Stardust mission, 105, 107, 141, 143, 236
Stellar evolution, 192
Sticking efficiency, 50

Subgroups, 6–8, 97, 148, 249
Supernova, 2, 3, 5, 26, 100, 102, 165,
    188, 197–199, 225, 227, 234

Total visual extinction, 67, 82, 225

UIR bands, 2, 186
Ultraluminous infrared galaxy, 216
Ultraviolet absorption bump, 34
Unidentified infrared features, 2, 31

Vega-type stars, 191
Very small grains, 52, 86, 92, 211,
    215, 217, 218
Vibrational frequency, 7
Virgo and Coma clusters, 218, 219,
    231

Young stellar objects, 52, 187